ANALOG VLSI INTEGRATION OF
MASSIVE PARALLEL SIGNAL PROCESSING SYSTEMS

ANALOG VLSI INTEGRATION OF MASSIVE PARALLEL SIGNAL PROCESSING SYSTEMS

by

PETER KINGET
Katholieke Universiteit Leuven,
Heverleee, Belgium

and

MICHIEL STEYAERT
Katholieke Universiteit Leuven,
Heverlee, Belgium

Kluwer Academic Publishers
BOSTON / DORDRECHT / LONDON

A C.I.P. Catalogue record for this book is available from the Library of Congress

ISBN 0-7923-9823-8

Published by Kluwer Academic Publishers,
P.O. Box 17, 3300 AA Dordrecht, The Netherlands.

Kluwer Academic Publishers incorporates
the publishing programmes of
D. Reidel, Martinus Nijhoff, Dr W. Junk and MTP Press.

Sold and distributed in the U.S.A. and Canada
by Kluwer Academic Publishers,
101 Philip Drive, Norwell, MA 02061, U.S.A.

In all other countries, sold and distributed
by Kluwer Academic Publishers Group,
P.O. Box 322, 3300 AH Dordrecht, The Netherlands.

Printed on acid-free paper

All Rights Reserved
© 1997 Kluwer Academic Publishers
No part of the material protected by this copyright notice may be reproduced or
utilized in any form or by any means, electronic or mechanical,
including photocopying, recording or by any information storage and
retrieval system, without written permission from the copyright owner.

Printed in the Netherlands

Contents

List of Symbols, Notations and Abbreviations		vii
Preface		xi
1.	ANALOG PARALLEL SIGNAL PROCESSING	1
	1.1 Introduction: Biological information processing systems vs artificial information processing systems.	1
	1.2 Artificial Neural Networks	3
	1.3 Implementation of artificial neural networks	4
	1.4 Analog VLSI and parallel signal processing	10
	1.5 Cellular Neural Networks	12
	1.6 Conclusions and Scope of this Research	19
2.	IMPLICATIONS OF MISMATCH ON ANALOG VLSI	21
	2.1 Introduction	21
	2.2 Modeling and characterization of transistor mismatch	23
	2.3 Implications of mismatch on transistor behavior	35
	2.4 Implications of mismatch on elementary stages	40
	2.5 Implications of mismatch on analog system performance	60
	2.6 Techniques to reduce impact of mismatch	73
	2.7 Link with Harmonic distortion	76
	2.8 Implications for analog parallel signal processing systems	78
	2.9 Conclusions	80
3.	IMPLEMENTATION ORIENTED THEORY FOR CNN'S	83
	3.1 Introduction	83
	3.2 Influences of VLSI imperfections on network operation	86
	3.3 Random static errors	87
	3.4 Generation of Accuracy Specifications	99
	3.5 Random dynamical errors	101
	3.6 Systematic static errors	108
	3.7 Systematic dynamic errors	111

3.8	Noise in CNN's	114
3.9	CNN's as resistive grids	114
3.10	Robust Template design	115
3.11	Conclusions	117

4. VLSI IMPLEMENTATION OF CNN'S — 121

4.1	Introduction	121
4.2	Analog VLSI implementation of computation operations	122
4.3	Programmable weighting of signals	124
4.4	Input-Output (I/O) circuits	141
4.5	VLSI cell architecture for CNN-cell	147
4.6	4x4 fully programmable CNN prototype chip	157
4.7	20x20 Analog parallel array processor	170
4.8	Performance evaluation with other implementations	192
4.9	Conclusions	195

5. GENERAL CONCLUSIONS — 197

Appendices — 202

A– MOS transistor models — 203

Bibliography — 211

Index — 223

List of Symbols, Notations and Abbreviations

List of Abbreviations

This list describes the symbols and abbreviations used throughout the text. The physical constants and used transistor parameters are described in appendix A.

2D	two-dimensional;
A/D	analog to digital converter;
ANN	artificial neural network;
APAP	analog parallel array processor;
BiCMOS	bipolar complimentary metal-oxide-semiconductor;
BW	bandwidth;
CMOS	complementary metal-oxide-semiconductor;
CNN	cellular neural network;
D/A	digital to analog converter;
DSP	digital signal processor;
GBW	gain-bandwidth product;
IC	integrated circuit;
I/O	input/output;

nMOS	n-type complementary metal-oxide-semiconductor;
OTA	operational transconductance amplifier;
pMOS	p-type complementary metal-oxide-semiconductor;
PC	personal computer;
PCB	printed circuit board;
VLSI	very large scale integration.

Notations

The following notations for the subscripts of voltage and current signals are used to indicate their total, DC, AC, ... values [Lak 94]:

I_{OUT}	DC or average value of a current signal;
i_{out}	instantaneous value of AC component of a current signal;
i_{OUT}	total instantaneous value of a current signal so $i_{OUT} = I_{OUT} + i_{out}$;
I_{out}	amplitude of the AC component of a current signal in steady state;

Symbols

The following symbol conventions are used in the circuit schematics throughout this text:

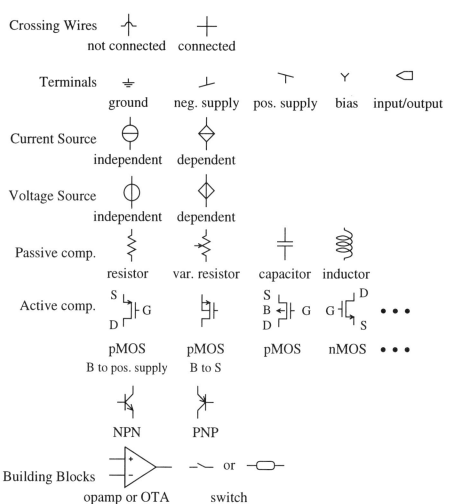

Preface

When comparing conventional computing architectures to the architectures of biological neural systems, we find several striking differences. Conventional computers use a low number of high performance computing elements that are programmed with algorithms to perform tasks in a time sequenced way; they are very successful in administrative applications, in scientific simulations, and in certain signal processing applications. However, the biological systems still significantly outperform conventional computers in perception tasks, sensory data processing and motory control. Biological systems use a completely different computing paradigm: a massive network of simple processors that are (adaptively) interconnected and operate in parallel. Exactly this massively parallel processing seems the key aspect to their success.

On the other hand the development of VLSI technologies provide us with technological means to implement very complicated systems on a silicon die. Especially analog VLSI circuits in standard digital technologies open the way for the implementation of massively parallel analog signal processing systems for sensory signal processing applications and for perception tasks. In chapter 1 the motivations behind the emergence of the analog VLSI of massively parallel systems is discussed in detail together with the capabilities and limitations of VLSI technologies and the required research and developments.

Analog parallel signal processing drives for the development of very compact, high speed and low power circuits. An important technological limitation in the reduction of the size of circuits and the improvement of the speed and power consumption performance is the device inaccuracies or device mismatch. In chapter 2 device mismatch and its impact on analog circuit and system design is discussed in detail. The accurate quantitative modeling and characterization of transistor mismatch is an essential prerequisite for the successful design and optimization of high performance analog systems and is therefore discussed first. The second part of chapter 2 deals with the impact of transistor mismatch on the optimal design of basic building blocks towards an optimal total performance including speed, accuracy, gain and power consumption. This knowledge is then extended and leads to derivation of the fundamental limitation that is put onto the different specifications of analog systems and building blocks by device mismatch. These limitations are compared with other ones like noise and distortion limitations.

Since technology limitations link the different performances of systems and basically put fundamental limits on the best possible total performance of a system, good specifications for the building blocks are extremely important for a circuit designer. Especially in the design of analog parallel systems a good understanding of the impact of technology limitations on the system performance is necessary for the circuit designer to come up with compact efficient VLSI implementations of the signal processing system. This calls for a joint effort in the development of the necessary theoretical methods to evaluate the impact of circuit limitations based on the experience from the circuit design. In chapter 3 this development of implementation-oriented theory is addressed for cellular neural networks. Dedicated theoretical methods are developed for the generation of good accuracy specifications for the building blocks dealing with the impact of random circuit errors so that a high circuit yield can be guaranteed. Also the impact of systematic errors on the correct operation of the system is discussed in detail.

Thanks to this theoretical background, efficient and compact analog hardware solutions for the implementation of computation functions are developed in chapter 4 in standard digital CMOS technologies. These circuits are then combined for the realization of a fully programmable CNN system for real-time sensor signal processing. The experimental results for two chip implementations together with the necessary sensor and computer interfaces to build the full real-time 2D sensor signal processing system are also reported in chapter 4.

Finally we have to express our gratitude to all persons who have contributed directly or indirectly to the realization of this book. We would like to thank Prof. H. De Man, Prof. P. Jespers, Prof. W. Sansen and Prof. J. Vandewalle for proofreading the manuscript and making many useful comments. We would also like to thank J. Bastos and R. Roovers for the fruitful collaboration on the transistor mismatch research and the other members of the ESAT-MICAS research group for the stimulating discussions. We would also like to express our gratitude towards the Belgian National Fund for Scientific Research (NFWO) and to Mietec-Alcatel for logistic support of the research.

Last but not least we thank our families for their year-long support and their patience which have significantly contributed to the realization of this research and this book.

Peter Kinget
Michiel Steyaert

Department of Electrical Engineering - ESAT MICAS
Katholieke Universiteit Leuven
Leuven, Belgium, 1996

1 ANALOG PARALLEL SIGNAL PROCESSING

1.1 INTRODUCTION: BIOLOGICAL INFORMATION PROCESSING SYSTEMS VS ARTIFICIAL INFORMATION PROCESSING SYSTEMS.

When the capabilities of present-day artificial information processing systems are compared to the capabilities of biological systems a few striking conclusions have to be drawn. Our present-day computers are mainly built with serial processors, which can process only one instruction on one data element at a time but they achieve a very high clock speed and accuracy. In this way they can perform a whole set of data-manipulation operations on very large data sets at very high speed. In order to solve a given problem an explicit algorithm or solution strategy has to be conceived, that when correctly programmed, yields a correct result after execution. In administrative applications, in scientific simulations, and for certain signal processing applications this computing power has been applied very successfully and has supported much of the technological and scientific progress of the last decennia. But in several application fields, however, the impact of the introduction of computers has not led to the expected solution of problems.

Biological information processing systems have a remarkable performance in real-time sensory data processing, in perception tasks, and in motory control applications. Even small insects e.g. can have exceptionally fast reactions to avoid dangerous situations or can locate (hidden) food sources very efficiently. This is partly due to the extraordinary sensitive sensors these animals have, but also in large part thanks to the very successful data processing they perform in their nervous system. Artificial robots are still very clumsy at moving around on unknown terrains with moving obstacles or no computer up to date is capable of understanding fluently human speech or can easily interpret images and recognize people, patterns or objects. Clearly at these perception tasks the biological information processing systems are much more efficient. On the other hand, the serial computer beats by far the biological information processing systems in mathematical computations at high speed.

Many biological information processing systems also are adaptive to their environment and improve their performance considerably over time through learning from examples. Even if some part of the system starts malfunctioning at a certain point in time, the adaptivity enables the system to correct after a certain learning period.

The operation principles of biological neural systems have been studied for many years in order to better understand their functioning; eventually we would like to apply this knowledge to build new types of artificial information processing systems. In very general terms, biological neural systems are typically made of a very large number of processing elements, called *neurons*, that operate at a relatively low speed and have a relatively low accuracy in their computation. These neurons are interconnected with connections with an adaptable strength, called *synapses*. Depending on the location and the function in the system, the interconnections can be to the near neighbors only, like e.g. in the retina, or can be very dense (up to many synapses per neuron). The neurons basically sum the information coming in on the synapses, compare it with a threshold and generate a non-linear output response. In contrast to the programmed computing paradigm of conventional computers, where the algorithm is explicitly described and stored in memory, in a neurocomputing system, the information processing algorithm is stored/captured in the strength of the connections between the neurons and thus also the interconnection pattern between the neurons. This distributed storage provides extra robustness to the system; if one of the connections fails, the whole system will not fail completely, but merely shows a lower performance and degrades gracefully. Moreover through the massive parallellism used for the computations, there is inherently a high level of redundancy in the system and the failure of single elements sometimes even does not result in any distinguishable system performance degradation.

We summarize that conventional computer systems and biological information processing systems have a completely different architecture: computers have a single very performant processing element whose operations are sequenced in time to obtain a result; biological neural networks contain a high number of simple neurons (processors), interconnected with variable strength connections, that obtain their performance from a *massive parallellism* in the operations. The success of these architectures is clearly different for different types of tasks or application fields and this difference in effectiveness is probably linked to the difference in architecture and operation principle.

1.2 ARTIFICIAL NEURAL NETWORKS

The idea of trying to apply the computation principles of biological neural systems in artificial systems exists for a few decennia and has been denoted artificial neural network or neurocomputing research. The earliest work dates from the 1940's when McCulloch and Pitts [McCu43] introduced their formal neuron model and showed it could perform computations; in the same period Hebb [Heb 49] developed the first models for learning and adaptivity. These results motivated many other researchers to join the field and inspired already the construction of the first neurocomputers in the 50's and 60's. These early accomplishments created very high expectations but the field received a severe set-back in 1969 when Minsky and Papert [Min 69] mathematically proved the limited capabilities of the models available at that time. During the 70's until the beginning of the 80's, the field lost much of its wide-spread interest; however, during these silent years some very firm mathematical and biological foundations for the field where formed. They supported the renewed interest in the mid 80's which was triggered by the papers of Hopfield [Hop 84], the interest of the American DARPA agency [DAR 88] and the publication of the 'Parallel Distributed Processing' volumes [Rum 88]. Moreover, the more powerful software tools and computers for simulation and the progress in the fabrication of integrated circuits offered the necessary technology for the first small applications of neurocomputing. A collection of the classic papers of neurocomputing through 1986 can be found in [And 88].

It must be recognized that the architecture of biological neural systems has been selected and refined through the struggle for the fittest during the very long evolution in nature. From an engineering point of view, it is probably not strictly necessary to build faithful copies of biological neural networks but artificial neural networks should exploit the same key principles in the context of the capabilities and the constraints of the technology we have available, which is very different from biology. For the understanding or studying of the operation of real biological neural networks, it is of course very helpful to faithfully imitate the functions of biological systems, to extend the experimental tool-set for researchers. They allow to develop models of the brain to study how the cognitive aspects of the brain could emerge. [Mah 91]

Many paradigms and architectures of artificial neural networks (ANN's) have been proposed. Two main classes exist. Multi-layer *feed-forward* NN's are built from several layers of neurons (typically 3 sometimes more); the neurons in the layers are connected by feed-forward connections only. These networks basically can be trained to perform any non-linear mapping from the input vector to the output vector. The most widely applied supervised learning algorithm is backpropagation; the network is presented with examples of correct input and output pairs; during learning the errors are propagated back into the hidden layers to update and improve the weights of the interconnections. A very important un-supervised learning law is used in the Kohonen-networks, which map multi-dimensional data into a compressed representation in their neurons through the process of self-organization. Many more learning laws exist and a detailed discussion can be found for instance in [Hec 90, and its references] or [DAR 88].

Feedback or recurrent NN are a second very important class of ANN's. They contain feed-back connections between the neurons so that dynamics become very important in the behavior of these networks. The networks are complex non-linear dynamical systems and their computation results can be equilibrium points or even limit cycles and more complex dynamical behavior [Thi 96]. This dynamical behavior can be used to build associative memories, solve optimization problems or perform signal processing tasks. Important examples of this type of networks are e.g. Hopfield networks and cellular neural networks.

Neural networks can be further classified in more detailed sub-classes, depending on the density or structure of their interconnections, the shape of their neuron non-linearities and many other properties. Many general textbooks are available that give an exhaustive description and classification of the field (see for instance in [Hec 90, and its references] or [DAR 88]).

1.3 IMPLEMENTATION OF ARTIFICIAL NEURAL NETWORKS

1.3.1 *Software Implementation*

For the implementation of ANN's many technologies are available with specific advantages and shortcomings. Software implementations of ANN's are very flexible; they can be reprogrammed and reconfigured very easily. Therefore they are very helpful in the exploration of the capabilities of ANN's. Software however is executed on classical computers so that the ANN operations are 'serialized' that the spatio-temporal parallellism of the ANN architecture is completely lost again. The simulation of large (dynamic) networks, requires very large simulation times and real-time processing applications are impossible. The volume of a software-implemented neural system is also much larger as networks directly implemented in hardware. As such software implementations are mainly useful for the exploration of the capabilities of ANN's but for real-life applications a hardware implementation is necessary.

1.3.2 Hardware Implementation

Hardware implementations can conserve the massively parallellism of ANN's and their high operation speed. The design of hardware is however more demanding and hardware can not always so easily be modified or updated. For the hardware implementation, analog or digital technology can be applied. In all areas of information processing, there are debates as to whether analog or digital signal processing is the more advantageous approach. Traditionally a digital approach has offered greater flexibility and high accuracy. The analog approach offers speed, size and power advantages.

Analog or Digital Hardware. The design of *digital systems* can be very well formalized and is strongly supported by computer tools. Digital systems are very robust and can easily attain very high accuracies. Digital systems obtain these advantages from two important paradigms. Digital systems only consider two valid signal levels, high (1) or low (0). More accurate signals are represented in binary format with a number of 1/0 or bits in parallel. This technique makes digital circuits very robust since extremely large noise or other interfering signals must be present to misinterpret a 1 as a 0 or vice-versa. The accuracy is easily scalable by adding extra bits in the signal representation. Secondly, most digital systems use time-discrete signals and synchronous or clocked circuits[†]. The signals are only valid at given points in time indicated by the clock signal. From clock pulse to clock pulse, the circuits compute the new value for the signals. These two paradigms allow digital systems to be accurately modeled with mathematical techniques; they can be very well simulated and also the synthesis of digital systems is very successful. But digital circuits need many transistors to process one bit so that a large circuit area is required and a lot of power is consumed. Due to the clocked nature of the systems, their speed is inherently lower.

Analog circuit implementations represent the signals as physical quantities, like e.g. charge, current, voltage, magnetic flux, time duration or frequency. These signals are continuous in value and most are continuous in time. Therefore analog implementations are typically an order of magnitude faster as dig-

---

[†]Nowadays asynchronous digital systems attract a lot of attention again. They do not use a global clock to control the system operation, but they still have to restrict the speed of the signals to ensure a correct system operation.

ital implementations. During each clock period a digital system has to settle and the maximal signal frequency must be at least two times smaller than the clock frequency to avoid signal aliasing (Nyquist Theorem). An analog system can exploit the full available technological bandwidth as the highest processing speed. The maximal signal frequency in analog circuits is typically an order of magnitude higher than in digital systems. In all signal processing systems, the highest speed signal operations are still performed by analog circuits.

In analog circuits, all properties of circuit devices can be fully exploited for the implementation of signal processing operations. This yields very compact and also very power efficient systems, since more than one bit can be processed per transistor. Analog circuits and systems are however, much harder to design; they are susceptible to temperature variations, process parameter variations, thermal noise, device inaccuracies and clock-feed through. If very high accuracies are required, all these parasitic effects have to be suppressed, which requires extra circuitry and results in extra area consumption, lower speed and more power consumption. For very high accuracies the impact of these parasitics becomes very important. In chapter 2 we will show that for applications with a low signal to noise requirement on the building blocks, analog implementations are much more efficient than digital implementations.

Because of these clear advantages of analog implementations for the implementation of ANN's, a lot of research has been conducted in this area. The availability of VLSI technologies, which enable the realization of enormous numbers of devices on a single chip, allow the realization of full systems on a chip. This technology opens new opportunities for the implementation of massively parallel systems and ANN's of considerably large size so that they can benefit from their parallel architecture. The limitations of analog circuits however, urge for a good understanding of the implemented systems and their sensitivity to hardware limitations and inaccuracies [Ver 92].

Interconnections. An important technological limitation faced in the design of neural network hardware is the high number of interconnections between the cells or neurons that are used in many network types. Chip fabrication technologies are planar technologies and allow for 2 to 7 metal interconnection levels[†]. This is much more restricted than the 3D interconnection space available in biological networks. More and more solutions for this problem emerge; the architecture of the networks is tuned to the capabilities of the VLSI technology and the limited interconnection level is exchanged for the very high available bandwidth. Cellular neural networks e.g. only require local connections and still perform global operations through the propagation of information. Also

---

[†]the number of interconnection levels increases for fine line-width technologies; the area consumption by wiring and interconnections in digital systems realized in these technologies becomes so dominant that further device down-scaling is only useful if more interconnection levels become available.

multi-chip solutions for large networks are being developed [VdS 92, Hei 93], together with efficient communication strategies that allow the multiplexing of a large number of connections over a few global bus wires [Mor 94, and its references].

Analog Memory and Adaptivity. Many of the ANN paradigms require the adaptation of the system to the environment through a learning process. At the hardware level this requires a memory for every connection to store its value and an adaptable synapse circuit. In present standard VLSI technologies the lack of a good analog memory cell is one of the major limitations for the development of trainable ANN chips [Vit 90a]. The available dynamic analog memory cells (e.g. a sample and hold function using charge storage on a capacitor) have only short retention times, and require an update or refresh of their value from another (external) static memory, which creates an interconnection or communication bandwidth bottleneck. Floating gate type of memories have very long retention times but require very long programming times and very high programming voltages and efficient types are only available in special EEPROM technologies. Hybrid solutions which combine digital 1-bit static memories with analog techniques to obtain an internal D/A conversion and to represent the signals in an analog format, are a feasible but very bulky solution. Some medium term memory solutions based on self-refreshing analog memory cells using discrete levels [Mac 93] are becoming available and promise a rather efficient solution especially for systems where a continuous learning mode is possible so that an automatic refresh is obtained. However in more classical adaptive systems with a limited number of adaptive coefficients these memory cell designs are sufficient and open the way to the analog implementation of high speed, low power and compact adaptive systems.

Applying learning algorithms directly in hardware can yield very important advantages. Robust learning laws can compensate for the circuit deficiencies; the training acts as feedback to compensate for variations in the characteristics of individual devices in the network. This further reduces the specifications on the hardware so that simpler, more compact and power efficient analog hardware can be designed. At this point however, the lack of a compact flexible analog memory cell is strongly limiting the inclusion of on-chip learning systems.

VLSI Implementation Technology. For the design and implementation of analog circuits three mainstream types of technologies are available. Chips fabricated in *CMOS technologies* form the bulk of the semiconductor industry since they represent the large market of digital systems, microprocessors and digital signal processors. Also a lot of analog systems and nowadays more and more mixed signal (analog and digital circuits combined on one die) systems are realized in CMOS. The high end analog circuits are designed and fabricated in *analog bipolar technologies* which offer the highest technology speeds and the best power efficiency. *BiCMOS technologies* try to combine the advantages of both technologies at the cost of a more complex and thus more expensive fabrication process. For a designer however, this technology offers the highest design flexibility and this technology is very popular for the design of single chip mixed signal systems.

The most important requirement for the design and fabrication of massively parallel analog computation systems is the availability of compact devices and high density interconnect technologies. This requirement is much more related to the technological demands of digital systems than of classical analog signal processing circuits. Also in digital circuits mainly the interconnection density determines the complexity of the system that can be realized on a chip die. In classical analog building blocks, on the contrary, the performance is mainly determined by the quality of the active and passive devices.

The development of CMOS technologies is completely driven by the requirements of digital systems. In table 1.1 the roadmap projection for the development of the digital microprocessor CMOS technologies is summarized [Gep 96]. As the line-widths keep shrinking, and the number of devices on the chip keep increasing, full systems will be housed on a single die; and these trends are projected to endure well into the next decade. The number of interconnection levels and the number of I/O pins per chip also keep increasing in future technologies which will result in a very high interconnection density. The complexity of the chips is not only increasing thanks to the down-scaling of the devices and the interconnects but the also the size of dies keeps expanding so that more functions can be packed on a single die. Moreover, thanks to the down-scaling of devices and interconnects the clock speeds of the digital systems will attain the GHz range in the next decade.

It is clear that this technology evolution will make available an enormous potential for the realization of analog parallel systems. But the analog circuitry must be kept compatible with the more restricted design flexibility of these digital technologies. This requires a good understanding of the impact of the technology limitations on the correct operation of analog parallel signal processing systems; already at the system level these constraints have to be taken into account in the design and optimization of these systems.

Of course the analog BiCMOS and bipolar technologies will also scale further in the future. Due to the complexity of the processing in BiCMOS, however, the device feature sizes and the interconnect density is typically one or two generations behind the digital CMOS standard. For the implementation of

Table 1.1. Roadmap projections for the microprocessors CMOS technology [Gep 96].

Year	Smallest feature [μm]	Chip size [mm^2]	Millions of transistors per cm^2	Clock [MHz]	Wiring levels	I/O per chip
1995	0.35	250	4	300	4-5	900
1998	0.25	300	7	450	5	1350
2001	0.18	360	13	600	5-6	2000
2004	0.13	430	25	800	6	2600
2007	0.10	520	50	1000	6-7	3600
2010	0.07	620	90	1100	7-8	4800

parallel systems using a standard digital CMOS technology clearly allows the realization of more complex systems.

The development of bipolar technologies is completely driven by the requirements of high speed, high performance analog building blocks for e.g. communications applications. The improvements of one generation to the next is mainly in the cut-off frequencies of the active devices. Massively parallel analog systems, however, do not obtain their performance from the speed of the single devices but rather from the parallel operation of a high number of interconnected devices; therefore the availability of compact devices and interconnects is more desirable than the availability of very high performance devices. Bipolar technologies are thus not suited for the implementation of massively parallel analog signal processing systems.

Artificial neural networks or massively parallel signal processing systems have risen very high expectations. However, their implementation is at this time still not at a level to fulfill these expectations. The lack of analog memories is a very important limitation for the implementation of trainable ANN systems. But trainability is not the only strength of ANN's. The massively parallel architecture is another very important feature of artificial and biological neural systems and the currently available and certainly the future standard digital CMOS technologies will provide us with the tools to implement this kind of massively parallel architectures, as long as their architectures take into account the limited 2D interconnection strategy of the planar IC technologies and the restricted device flexibility.

1.4 ANALOG VLSI AND PARALLEL SIGNAL PROCESSING

Sensor Signal (Pre)Processing. The observations presented in the previous paragraphs are backed by the trends in the analog VLSI hardware research. In recent years a lot of research has emerged in the analog VLSI field on the implementation of analog parallel signal processing or computation devices[Mea 89, Vit 90b, Vit 94]. These systems basically use parallel architectures but do not include detailed learning strategies. Many have however a certain degree of adaptive behavior included in the system. These systems are used to process sensor signals [Kob 90, Sta 91, Gru 91, Yu 92, Esp 94b, Che 95, Arr 96, Veni96] or are sensor-driven networks like artificial retina's, motion sensors or cochlea's [Mea 89]. They represent a new system paradigm for the processing of sensor signals.

In the classical sensor signal interface and processing system, depicted in figure 1.1(a) [Grat91], the analog circuitry is used for capturing, amplification, filtering and other conditioning operations. All feature extraction operations are done in the digital domain after an A/D conversion of the conditioned sensor signal. For the processing of 2D sensor signals, the throughput requirements on the A/D conversion and the digital signal processing hardware become very high if a real-time operation is to be obtained. Moreover, many of the required processing tasks in perception systems require only a low accuracy and can thus better be implemented in analog circuits.

This has led to the new architecture of figure 1.1(b). The low-level signal processing operations, i.e. mainly the feature extraction tasks, are merged with the classical analog interface electronics. Especially for the processing of 2D sensor signals, from image or tactile sensors for instance, this architecture has important advantages. It exploits the possibilities of integrating a simple processor per pixel so that a very high throughput can be achieved with a limited power consumption. Only the relevant features, which represent a much lower signal bandwidth, have to be converted to the digital domain; the higher level processing and classification tasks can then be executed by a much lower performance digital processor. The combination of the sensors and the parallel analog signal processing circuits yields a smart sensor device.

As such, these systems have been inspired by the neural network research, but they clearly differ in their aim. They try to exploit the advantages of analog VLSI design and parallel processing in the best possible way, without trying to directly implement or mimic a biological system. At this point they are the most successful representations of the possible advantages of analog parallel signal processing systems. Their success is due to the relatively small or simple paradigms that are used but that enable a thorough understanding of the system behavior so that the hardware requirements to obtain a high

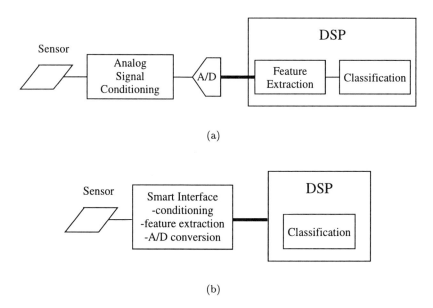

Figure 1.1. (a) The classical sensor interface system architecture. (b) Sensor interface using massively parallel analog signal processing for feature extraction.

performant signal processing system can be derived. We can conclude that the analog implementation of neural networks and parallel signal processing systems is a matter of bringing together application oriented problem analysis with the circuit architecture design and optimization. A separate effort is one of these domains is not thought effective.

1.5 CELLULAR NEURAL NETWORKS

Many of the theoretical and hardware techniques to be presented in this work are applied for cellular neural neural networks. In this section the cellular neural network paradigm is introduced and a few important properties of cellular neural networks are summarized [Chu 88b, Chu 88a].

Definition. A cellular neural network consists of an uniform 2D array of analog nonlinear dynamic computing cells, which are only connected to their nearest neighbors, as is depicted in figure 1.2. The interconnection strengths or connection weights are spatially invariant. They are organized in *weight templates*, which are thus identical for all cells. The network contains no adaptive behavior or learning, so it is functionally related to a resistive grid. Resistive grids have been known for a long time and have been applied for several smoothing or enhancing (when negative resistors are used) operations in image processing for example. They are however static systems, and solve a set of linear or non-linear equations. Resistive grids can be casted as a special type of CNN [Shi 92]. CNN's, however, are dynamical systems and their behavior is described by non-linear differential equations. By using dynamical effects they can compute more complex and global operations so that they have more powerfull capabilities.

All cells in the CNN operate in parallel and in continuous time; one computing cell per pixel in the 2D input signal is allocated which results in very high signal processing speeds. Although the cells are only locally connected, the network is able to perform global operations on the 2D inputs like e.g. connected component detection, hole-filling and shadow making. This is possible through the propagation of information through the network from cell to cell; in this way, the missing global connections are replaced by a time-multiplexing of the connections and the time-propagation of the information, thanks to the dynamical nature of the networks.

State Equations. Each cell in the network has a state x_i, an output y_i and an input u_i. The evolution and the dynamics of the state of cell i is described

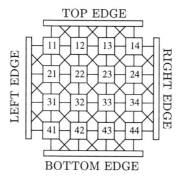

Figure 1.2. The structure of a 4x4 cellular neural network with nearest neighbors connections. The boxes represent the cells and the lines represent the connections. The cells are numbered like the elements of a matrix.

by the non-linear differential equation:

$$\frac{dx_i}{dt} = -x_i + I + \sum_{c \epsilon N_r(i)} A_c \cdot y_c + \sum_{c \epsilon N_r(i)} B_c \cdot u_c \quad (1.1)$$

and

$$y_i = f(x_i) = \frac{1}{2}(|x_i + 1| - |x_i - 1|) \quad (1.2)$$

$N_r(i)$ is the r-neighborhood of the cell i and it contains all cells within a radius r; the 1-neighborhood e.g. contains the 8 nearest neighbors and the cell itself, which is the most used connection scheme for CNN's. The output nonlinearity $f(x_i)$ is a piecewise linear non-linearity; f is linear in the unit range $[-1, 1]$, and outside the unit range the output saturates to $+1$ for positive state values and to -1 for negative state values. The range $[-1, 1]$ of state values is also called the *linear state region* whereas the ranges $[-\infty, -1]$ and $[1, \infty]$ are called the *saturation state regions*.

The block diagram of the cell in figure 1.3 implements equation (1.1). The cell sums the incoming signals from the neighbors (and itself) and the constant bias of the I-template and integrates them to compute its state evolution. At all times it sends two signals to each of its neighbors: one signal is its output multiplied by a weight from the A-template; the second signal is its input multiplied by a weight from the B-template. The A template connections provide an output feedback mechanism whereas the B template connections serve as an input control mechanism. These weight patterns or templates are identical for all cells in the network. For the standard nearest neighbor connections, the A

14   ANALOG VLSI INTEGRATION OF MASSIVE PARALLEL SYSTEMS

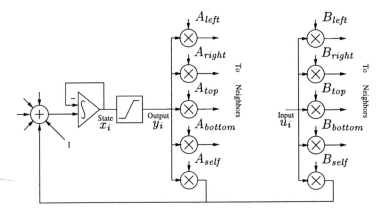

Figure 1.3.   Block diagram of a cell of a cellular neural network.

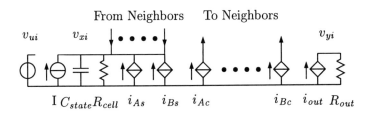

Figure 1.4. Schematic for an electrical CNN cell implementation: $i_{Ac} = A_c \cdot v_{yi}$, $i_{Bc} = B_c \cdot v_{ui}$, and $i_{OUT} = 1/2R_{out}(|v_{xi} + 1| - |v_{xi} - 1|)$.

and B templates can be represented as 3x3 matrices as is illustrated in table 1.2. The three templates A, B and I, consist together of 19 weight parameters and define the operation and the function executed by the network.

In figure 1.4 a schematic of an electrical implementation of a CNN cell is represented. The node voltage v_{xi} across capacitor C_{state} is the state of the cell. The node voltage v_{ui} represents the input of the cell. The output is represented by the voltage v_{yi} generated across the output resistor R_{out} by a non-linear voltage controlled current source (i_{OUT}) that is controlled by the state voltage. The linear capacitor C_{state} and the linear resistor R_{cell} form a lossy current integrator. The linear voltage controlled current sources (i_{Ac} and i_{Bc}) generate current signals that are sent to the states of the neighboring cells and the linear dependent sources i_{As} and i_{Bs} provide the self-feed-back connection and the input feed-forward connection. The independent current source I adds the constant I-template.

Operation Cycle. Two different modes of operation exist for a CNN: input-driven or autonomous. In the *autonomous mode*, the input nodes of the cells are not used and weights in the B-template are zero. The 2D input signal is loaded as the initial state of the network, and the result of the computation is the final equilibrium state and the corresponding outputs of the network. Connected component detection or averaging are two templates of this type (see table 1.2).

In the *input-driven mode*, the pixels of 2D input signal are loaded on the input nodes of the network cells and the states are initialized with a fixed value. After the evolution of the network, the final equilibrium states and the outputs of the cells contain the result of the processing operation or the computation.

Range of State variable. For the design of a physical CNN implementation, the dynamic range of the different signals in the cells is very important. In [Chu 88b] the state x_i of the cells is shown to be bounded at all times by:

$$x_{max} = 1 + |I| + \sum_{c \epsilon N_r(c)} (|A_c| + |B_c|) \quad (1.3)$$

under the condition that:

$$|x_i(0)| \leq 1 \quad (1.4)$$
$$|u_i| \leq 1$$

Condition (1.4) states that the magnitude of initial states of the cells and the inputs of the cells have to be smaller than one. The boundedness of the states expressed by (1.3) is due to the limiting action of the output non-linearity which limits the output y_i to $[-1, +1]$. For the VLSI implementation this result is very important; it provides the circuit designer with a specification on the signal swing on the cell state node. In table 1.2 the maximal swings of the state nodes given by (1.3) are listed for the different templates. The output node is bounded to $[-1, +1]$ by the output non-linearity and the output node is also constrained to $[-1, +1]$ by (1.4).

Stability. The function of a CNN for signal processing applications is the transformation of the input or the initial state into an output; a CNN must always converge to a constant steady state after a transient which has been initialized and/or driven by an input. The stability of dynamic nonlinear circuits is analyzed by defining a Lyapunov's function, which expresses the generalized energy present in the system. In [Chu 88b] a function is defined and shown to be bounded and monotone-decreasing under the assumption that the cell A template connections are *symmetric*. The properties of this Lyapunov's function imply that the states of the network will always evolve towards a constant

DC equilibrium value. Moreover, if the self feedback entry in the A template satisfies:

$$A_{self} > 1 \qquad (1.5)$$

each cell state settles at a stable equilibrium point with a magnitude greater than 1 so that the outputs of the network will be binary +1 or -1. This can be understood by the fact that (1.5) ensures that in the region $[-1, +1]$ a minimum net amount of positive feedback is present, that compensates the negative feedback from the loss term in (1.1), so that no stable state equilibrium points exist in between -1 and 1.

The symmetry or reciprocity in the A template connections is a strong condition. In VLSI implementations there will always be a slight difference between the connection weights due to device inaccuracies. Moreover certain templates explicitly use non-reciprocal connection weights. Stability for large classes of non-reciprocal CNN's has been demonstrated in [Chu 90, Chu 92, Guz 93], under much wider parameter conditions.

CNN template library. A large collection of templates is available for 2D linear and non-linear signal processing operations and for feature extraction and can be found in [Ros 94]. In table 1.2 a few templates are collected which are often used in this work.

Programmable CNN's Requirements. The function or operation executed by the CNN on the input signals is dependent on the A, B, and I template values. Since all cells are controlled by the same templates, only 19 weight values for a CNN with nearest neighbor connections, must be stored and changed to make the CNN function programmable or adaptive; the number of analog memories is only dependent on the interconnection topology and independent of the number of cells in the network. For the chip realization of a programmable or adaptive CNN, no memory problem thus exists since hybrid memory cells can be used.

A programmable CNN chip can be considered as an *analog parallel array processor*, with the templates as instructions; these instructions can be combined to form complete signal processing algorithms. The analog cells operate in parallel and the array structure is ideally suited for the processing of 2D input signals. In order to be able to execute the total template collection [Ros 94] for the standard CNN model, A and B template weight factors from $\pm 1/4$ to ± 4, and I template values up to ± 10 are required. Also a weight factor of 0 must be programmable to remove certain connections between the cells. The maximal value for the state range that occurs in the template collection is

Table 1.2. A small selection of typical CNN templates from [Ros 94].

Template-name	A	B	I	x_{max}
INFORMATION PROPAGATING TEMPLATES				
HORIZ. CONNECTED COMPONENT DETECTOR (CoCoD)	$\begin{bmatrix} 0 \\ 1 \quad 2 \quad -1 \\ 0 \end{bmatrix}$	$\begin{bmatrix} 0 \\ 0 \quad 0 \quad 0 \\ 0 \end{bmatrix}$	0	5
HOLE FILLING (HOLE_MOD)	$\begin{bmatrix} 1 \\ 1 \quad 2 \quad 1 \\ 1 \end{bmatrix}$	$\begin{bmatrix} 0 \\ 0 \quad 4 \quad 0 \\ 0 \end{bmatrix}$	0	11
SHADOW GENERATION	$\begin{bmatrix} 0 \\ 2 \quad 2 \quad 0 \\ 0 \end{bmatrix}$	$\begin{bmatrix} 0 \\ 0 \quad 2 \quad 0 \\ 0 \end{bmatrix}$	0	7
NON-PROPAGATING TEMPLATES				
EDGE DETECTION	$\begin{bmatrix} 0 \\ 0 \quad 2 \quad 0 \\ 0 \end{bmatrix}$	$\begin{bmatrix} -1 \\ -1 \quad 4 \quad -1 \\ -1 \end{bmatrix}$	-2	13
NOISE REMOVAL	$\begin{bmatrix} 1 \\ 1 \quad 2 \quad 1 \\ 1 \end{bmatrix}$	$\begin{bmatrix} 0 \\ 0 \quad 0 \quad 0 \\ 0 \end{bmatrix}$	0	7
HORIZONTAL PEEL	$\begin{bmatrix} 0 \\ 0 \quad 2 \quad 0 \\ 0 \end{bmatrix}$	$\begin{bmatrix} 0 \\ 3 \quad 3 \quad 0 \\ 0 \end{bmatrix}$	-5	14
LINEAR RESISTIVE GRID				
RESISTIVE GRID ($\lambda = \frac{R_{hor}}{R_{vert.}}$)	$\begin{bmatrix} 1/\lambda \\ 1/\lambda \quad -4/\lambda \quad 1/\lambda \\ 1/\lambda \end{bmatrix}$	$\begin{bmatrix} 0 \\ 0 \quad 1 \quad 0 \\ 0 \end{bmatrix}$	-2	1

about 16. This specifications are very important for the derivation of compact circuit implementations in chapter 4.

CNN Universal Machine. In [Ros 93] the CNN universal machine architecture is proposed which is an extension of the programmable CNN structure. It originates from the idea that by realizing programmable CNN implementations, several templates can be executed on the same hardware and combined in template sequences or algorithms; this in contrast to the dedicated fixed template implementations which require separate hardware for each function. Besides using an analog programmable CNN structure, the cell functionality is extended for the CNN universal machine architecture in mainly four ways:

- switches are added to realize a programmable cell and input/output (I/O) configuration;

- local logic circuitry is included to compute simple logic operations on the results of analog computations or previous logic operations;

- local analog memory cells store the results of previous analog computations;

- and local logic memories store the results of logic computations.

These local memory cells enable the reuse of previous computed results without requiring a global I/O operation so that very high throughputs can be attained.

On the system level the structure is extended with a global analog and logic programming unit which stores the analog and logic instruction sequences and controls the operation and configuration of the cells.

These extensions result in a programmable analog/logic array computer. The combination of the dual computing structure and the high throughput thanks to the local memory cells, results in possibilities for the development of algorithms combining the strengths of analog template processing and logic operations.

The main challenge for the VLSI realization of this system remains in the design of the analog building blocks for the programmable CNN subsystem. The extensions can be mainly realized through classical digital circuits and with a few fairly straightforward analog/digital interfaces and simple analog memory cells to store the -1/+1 output results. Although this structure opens nice application perspectives, we have concentrated in this work on the design of programmable CNN systems since they represent a key design and research challenge in the realization of massively parallel analog signal processing systems.

1.6 CONCLUSIONS AND SCOPE OF THIS RESEARCH

The ANN research has demonstrated the interesting potentials of massively parallel signal processing systems. The enormous integration capabilities of CMOS VLSI technologies and the developments in the design of compatible analog circuits in these technologies provide the fabrication tools for the realization of these systems. In the near future, these analog parallel signal processing systems show very promising applications in the field of real-time 2D sensor signal processing and perception-like applications.

In order to meet these expectations a joint research activity on two areas is required (see also e.g. [Ver 89]). On the system level, implementation oriented theory must be developed that deals with the impacts of implementation inaccuracies on the correct system operation and that provides the circuit designer with detailed building block specifications. On the hardware design level, this knowledge must then be applied to develop compact implementations for standard digital CMOS technologies with a high operation speed and a low power consumption.

In this work we demonstrate that through this joint effort successful chip implementations can be obtained that outperform classical digital implementations; we use programmable CNN systems as the demonstrator system for this research. On the other hand we also demonstrate that on the technology level, the impact of device mismatch puts fundamental boundaries on the ultimate performance that can be attained in chip implementations of these analog parallel signal processing systems and different other type of more traditional analog systems. Many of the results and circuits presented in this work are, however, also applicable for the VLSI design of other ANN systems e.g. [VdS 93, Lau 94] or of analog implementations of classical adaptive systems [Wid 85] and of the recently emerging continuous time adaptive systems [Deh 95, Lem 95]. In the long term, the implementation of biologically inspired systems including adaptive behavior and learning capabilities depends on the development of a compact, easy programmable analog memory cell with long retention times. The memory problem is not addressed in this work, but the design of the necessary adaptive multiplication circuits which are also required to implement adaptivity is covered in the presented circuit design results.

2 IMPLICATIONS OF TRANSISTOR MISMATCH ON ANALOG CIRCUIT DESIGN AND SYSTEM PERFORMANCE

2.1 INTRODUCTION

In a signal processing system several operations or computations are performed on a signal in different stages sequentially. Each of these operations have to emphasize a wanted component or property of the signal without adding too much unwanted extra components. These are due to the non-idealities of the circuit implementation compared to the specified operation. Circuit non-idealities can be divided in two groups: random and systematic errors.

The *random* errors are the result of the stochastic nature of many physical processes. The stochastic behavior of charge carriers in a conductor, for instance, results in various types of noise signals and the stochastic nature of the physical phenomena that take place during the fabrication of integrated circuits, results in a random variation of the properties of the fabricated on-chip devices and mismatches between identically designed devices.

The *systematic* errors occur because a typical circuit implementation only approximates an ideal signal processing operation to a limited extent. These errors are caused, for instance, by the non-linear operating characteristics of devices or by the influence of parasitics in the signal path or device structure.

The effect of these non-idealities can be of different kinds. The *noise* signals limit the minimal signal that can be processed with the system. Device *mismatch* limits the accuracy of the circuit behavior and again limits the minimal signal or energy that is required to execute meaningful signal operation functions. For linear systems, the non-linearities of devices generate *distortion* components of the signals or modulate unwanted 'noise' signals into the used signal band. This typically limits the maximal signal that can be processed correctly.

The circuit designer can reduce the effect of the distortion non-idealities by using small modulation indices for the bias signals; by using large device sizes the impact of mismatch is lowered and by using low impedance levels, the thermal noise signals are reduced. These measures have, however, very important consequences on the power consumption and operation speed of the system. Therefore the quality of a circuit realization is evaluated from the obtained accuracy, noise level or linearity relative to the used power and the speed of operation. The designer will try to achieve for a given speed the best performance with a minimal power consumption.

The fundamental impact of noise on the overall system performance has been studied extensively in literature see e.g. [Vit 90b] [Voo 93, and its references]. In this chapter we investigate the impact of transistor mismatch on the total performance of analog circuits and systems. First, we discuss the characterization and modeling of transistor mismatch and describe a new extraction method to derive the matching quality of sub-micron CMOS technologies. This quantitative model information is very important for the design of analog circuits since it allows the designer to accurately predict the accuracy performance. Furthermore, it forms the basis for the evaluation of the impact of transistor mismatch on the analog performance.

The implications of transistor mismatch on the design of basic analog building blocks is then discussed in detail in sections 2.3 and 2.4. The speed, accuracy and power consumption performances of analog circuits are linked due to the effect of mismatch on the circuit design; guidelines for the optimal design of circuits are derived. In section 2.5 we generalize these results and prove that mismatch puts a fundamental limitation on the maximal total performance of analog signal processing systems. The *Speed·Accuracy^2/Power* ratio is fixed by technological constants that express the matching quality of the technology. For circuit building blocks with high accuracy requirements thermal noise is considered as the limiting factor for performance improvement or power consumption reduction [Vit 94, Dij 94] but we show that the impact of transistor mismatch on the minimal power consumption is more important for present-day CMOS technologies than the impact of thermal noise for high speed analog circuits and massively parallel analog systems. The matching performance is technology dependent and the scaling of the circuit performance with the downscaling of the technology size is discussed in section 2.5.4. The techniques to reduce the impact of mismatch and their effect on the performance of systems is reviewed in section 2.6. The last section (2.8) treats the implications for the VLSI design of massively parallel analog systems and discusses the advantages of analog VLSI implementation over the digital VLSI implementation for massively parallel analog systems.

2.2 MODELING AND CHARACTERIZATION OF TRANSISTOR MISMATCH

2.2.1 What is transistor mismatch

Two identical designed devices on an integrated circuit have random differences in their behavior and show a certain level of random mismatch in the parameters which model their behavior. This mismatch is due to the stochastic nature of physical processes that are used to fabricate the device. In [Pel 89] the following definition for mismatch is given: *mismatch is the process that causes time-independent random variations in physical quantities of identically designed devices.*

2.2.2 Modeling of CMOS transistor mismatch

Mismatch in device parameters can be modeled by using different techniques. Several authors [Laks86, Shy 84, Miz 94] start from the physical background of the parameters to calculate and model the device mismatch dependence on technology parameters and device size. Or a black box approach can be used by supposing the statistical properties of the different mismatch generation processes and calculating their influence on different circuit parameters [Pel 89].

The mismatch of two CMOS identical transistors is characterized by the random variation of the difference in their threshold voltage V_{T0}, their body factor γ and their current factor β (the definitions of these parameters can be found in appendix A). For technologies with a minimal device size larger than typically 2 μm, a widely accepted and experimentally verified model [Pel 89, Bas 95, Pav 94] for these random variations is a normal distribution with mean equal to zero and a variance dependent on the gate-width W and gate-

Table 2.1. The matching proportionality constants for size dependence for different industrial CMOS processes. The parameter $(V_{GS} - V_T)_m$ is defined in (2.33) in section 2.3.

TECHNOLOGY λ_T	TYPE	A_{VT0} [mVμm]	A_β [%μm]	$(V_{GS} - V_T)_m$ [V]
2.5μm [Pel 89]	nMOS	30	2.3	2.6
	pMOS	35	3.2	2.2
1.2μm [Bas 95]	nMOS	21	1.8	2.3
	pMOS	25	4.2	1.2
0.7μm	nMOS	13	1.9	1.4
	pMOS	22	2.8	1.6

length L and the mutual distance D between the devices:

$$\sigma^2(\Delta V_{T0}) = \frac{A_{VT0}^2}{WL} + S_{V_{T0}}^2 D^2 \tag{2.1}$$

$$\sigma^2(\Delta \gamma) = \frac{A_\gamma^2}{WL} + S_\gamma^2 D^2 \tag{2.2}$$

$$\left(\frac{\sigma(\Delta \beta)}{\beta}\right)^2 = \frac{A_\beta^2}{WL} + S_\beta^2 D^2 \tag{2.3}$$

A_{VT0}, A_γ, A_β, $S_{V_{T0}}$, S_γ and S_β are process-dependent constants. In table 2.1 and 2.2 the proportionality constants for several processes are summarized. Experimental data show that the correlation between the V_{T0} and β mismatch is very low although both parameters depend on the oxide thickness[Pel 89, Bas 95].

The last two columns of table 2.2 contain corner distances at which the distance dependent term in the parameter mismatch becomes dominant over the size dependent term. The corner distances D_m is defined as the distance for which the mismatch due to the distance effect on a parameter p is equal to the mismatch due to the size dependence for a minimal size device ($W = \lambda_T$ and $L = \lambda_T$, where λ_T is the minimal size of the technology) and is calculated as:

$$D_m = \frac{A}{\lambda_T \cdot S} \tag{2.4}$$

For devices with an area of A_f times the minimal area the critical distance D_m is $\sqrt{A_f}$ times smaller. The obtained critical distances D_m for the present-

Table 2.2. The matching proportionality constants for distance dependence for different industrial CMOS processes.

TECHNOLOGY λ_T	TYPE	$S_{V_{T0}}$ [$\mu V/\mu m$]	S_β [ppm/μm]	D_{VTm} [mm]	$D_{\beta m}$ [mm]
2.5μm [Pel 89]	nMOS	4	2	3	5
	pMOS	4	2	3.5	13
1.2μm [Bas 95]	nMOS	0.3	3	58	5
	pMOS	0.6	5	35	12
0.7μm	nMOS	0.4	2	46	14
	pMOS	-	3	-	13

day processes are very large compared to the typical size of an analog circuit. Therefore the distance dependence of the parameter mismatch will be neglected in the discussion of the impact of transistor mismatch on analog circuit and system performance.

In equations (2.1), (2.2) and (2.3) the standard deviation of the difference of the parameters of two transistors is given. For a random variable Z defined as $Z = X - Y$ the variance is $\sigma^2(Z) = \sigma^2(X) + \sigma^2(Y)$. The following relations are thus obtained for the variance of the absolute parameters of a single transistor:

$$\sigma(V_{T0}) = \sqrt{2}\sigma(\Delta V_{T0}) \qquad (2.5)$$

$$\sigma(\gamma) = \sqrt{2}\sigma(\Delta \gamma) \qquad (2.6)$$

$$\left(\frac{\sigma(\beta)}{\beta}\right) = \sqrt{2}\left(\frac{\sigma(\Delta\beta)}{\beta}\right) \qquad (2.7)$$

2.2.3 Characterization of transistor mismatch

The matching behavior of transistors is very strongly dependent on the used IC technology. Therefore an in-house characterization procedure has been set-up [Bas 95]. A new direct extraction algorithm has been developed to extract the ΔV_{T0} and $\left(\frac{\Delta\beta}{\beta}\right)$ of a transistor pair from their measured relative current difference $\left(\frac{\Delta I_{DS}}{I_{DS}}\right)$ in saturation. The big advantage of measuring currents in saturation is the much lower sensitivity to parasitics in the set-up, which becomes more and more important for sub-micron and deep sub-micron technologies. Also, the model of V_{T0} mismatch of minimal sized devices in sub-micron technologies has been improved. In this paragraph a short overview is given; for more details the reader is referred to [Stey94] [Bas 95], [Bas 96c], [Bas 96b], and [Bas 96a].

A Test Circuits. Test-circuits are processed to experimentally check the validity of the models and to determine the proportionality constants of the size and distance dependence of transistor mismatch. In figure 2.1 a microphotograph of the nMOS test-chip for a 1.2 μm CMOS technology is presented. The test-chip contains a matrix of transistors. On a row identical transistors are spaced at different mutual distances to examine the mismatch spatial dependence. The different rows contain transistors with different sizes to determine the mismatch dependence on device size. All sources are connected to a common point. Transistors in the same row have their gates connected and transistors in the same column have their drains connected. Special attention has to be paid in the layout to obtain a very low resistance in the source path to eliminate systematic errors during the measurements; very wide source metal connections are used and can be clearly distinguished in figure 2.1. Two separate test-chips for the characterization of the nMOS transistors and pMOS transistors have been designed.

B Measurement set-up. The measurements are carried out on packaged test-circuits using a HP4062A Semiconductor Parametric Test System including a switch matrix and two voltage source units. The drain current of the different transistors in a row are accessed sequentially through a switch matrix which connects the drain of the transistor under test to the current meter; the drains of the other transistor in the row are left open. The switch-matrix connects the gate of the transistor under test to the gate voltage source and connects the gates of the other rows to ground. The transistors are biased in strong inversion by using gate voltages larger than V_T; the current is measured in saturation by applying a constant drain voltage larger than the maximal $(V_{GS} - V_T)$. The drain voltage is applied by using a *4 point technique* where two separate *sense* wires, which carry no current, monitor the drain voltage and the current flow is through two separate *force* wires. For the C12 technology, for instance, the gate voltage is swept in 26 steps from 0.75 V to 2.0 V with a constant drain voltage of 2.0 V. Since the different transistors are measured sequentially the DC repeatability of the DC gate voltage source must be larger than the smallest gate-voltage mismatch we want to measure. The repeatability of the source in our set-up was better than 6 digits which is more than sufficient.

For the extraction of the transistor mismatch we are interested in the current differences between the different transistor pairs. The current difference can only be obtained by measuring the currents separately and then subtracting the

IMPLICATIONS OF MISMATCH ON ANALOG VLSI 27

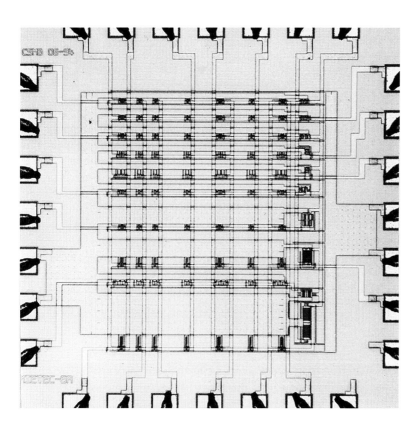

Figure 2.1. Micro-photograph of the mismatch characterization test-chip in a 1.2 μm CMOS technology.

current measurements. This procedure is very sensitive to errors, but it is the only way to obtain a measurement of the current difference of two transistors. To have an accurate estimation of the current difference, the individual currents have to be measured very accurately. The necessary relative accuracy on the the drain current measurement $\left(\frac{\sigma_m(I_{DS})}{I_{DS}}\right)$ is dependent on the relative current difference $\left(\frac{\Delta I}{I}\right)$ we want to measure and on the wanted relative accuracy for the current difference $\left(\frac{\sigma_m(\Delta I_{DS})}{\Delta I_{DS}}\right)$:

$$\left(\frac{\sigma_m(I_{DS})}{I_{DS}}\right) = \frac{1}{\sqrt{2}}\left(\frac{\Delta I}{I}\right)\left(\frac{\sigma_m(\Delta I_{DS})}{\Delta I_{DS}}\right) \tag{2.8}$$

The necessary number of digits in the current measurement must be larger as the number of digits we want in the current difference measurement plus the relative current difference we are measuring expressed in a number of digits. In our set-up a HP3457 multi-meter with a 5 to 7 1/2 digits resolution is used.

Using this procedure the current through all transistors in the array as function of the gate voltage is measured and stored.

C New Mismatch Parameter Extraction Technique. The relative current difference of a transistor pair is dependent on both the current factor β matching, the threshold voltage V_{T0} matching and the gate voltage; in section 2.3.2 the following dependence is derived for transistors biased in saturation:

$$\left(\frac{\Delta I_{DS}}{I_{DS}}\right) = \left(\frac{\Delta \beta}{\beta}\right) - \frac{2\Delta V_{T0}}{(V_{GS} - V_T)} \tag{2.9}$$

Since the current measurements are performed in saturation, the current of a transistor is in first order not dependent on the drain voltage but only on the gate-source voltage. Parasitic resistances in the drain path, from e.g. the switches in the switch-matrix or the current mirror, do not generate errors. However, differences in the parasitic resistors in the source path of the two devices under test generate systematic errors; therefore wide interconnects are used on the test chip in the source connections of the transistors.

First the threshold voltage is extracted for one of the transistors using a standard extraction technique [Bas 95]. Then the model of (2.9) is fitted to the relative current difference measurement using a linear least-squares algorithm. The mismatch parameters ΔV_{T0} and $\left(\frac{\Delta \beta}{\beta}\right)$ are obtained directly from the measured current difference.

Other techniques can be used to calculate the V_{T0} matching and β matching. In [Pel 89] the transistors are measured in the linear region and for each individual transistor the threshold voltage V_{T0} and the current factor β are extracted using classical parameter extraction algorithms. The parameter mismatch is then calculated by subtracting the parameters of the individual transistors.

Also a direct extraction technique can be derived from a more complex drain current model which includes a mobility reduction parameter θ. The extra mismatch parameter $\Delta\theta$ is however highly correlated to the $\left(\frac{\Delta\beta}{\beta}\right)$ and the overall modeling of the current mismatch becomes more complicated without improvements in accuracy [Bas 95].

We conclude that he new direct extraction procedure has two main advantages:

- The current measurements are performed in *saturation* so that parasitic resistances in the drain path do not generate errors. In sub-micron technologies where the current factor β is large due to the thinner oxide, the equivalent resistance of the transistors biased in the linear region becomes very small so that any parasitic series resistance in the current path gives rise to important errors. Thus especially for sub-micron technologies measuring in saturation region is a more robust technique. Moreover, the majority of the transistors are biased in saturation in analog design so that the mismatch parameters are obtained under realistic conditions.

- The parameter mismatches are extracted *directly* from the current difference measurements with a high accuracy; when first the absolute parameters are extracted and the mismatch parameters are calculated as a difference of absolute parameters, a very high accuracy in the absolute parameters is necessary to obtain accurate mismatch parameters - the same calculations (2.8) as for the current measurements can be used to calculate the necessary accuracy of the absolute parameters. A higher accuracy can be obtained in the individual absolute current measurements by using highly accurate measurement equipment than in the extraction of the absolute parameters and thus the presented direct technique gives more accurate results [Bas 95].

In this way a measurement of the ΔV_{T0} and $\left(\frac{\Delta\beta}{\beta}\right)$ is obtained for each of the transistor pairs, which all have different mutual distances and different gate areas, for every test-chip. From this experimental data the proportionality constants for the mismatch dependence on distance and on size can be extracted for the models in (2.1) and (2.3).

D Mismatch Dependence on Distance. For every row - containing transistors of the same size - the first transistor is used as a reference and the ΔV_{T0} and $\left(\frac{\Delta\beta}{\beta}\right)$ are extracted for the consecutive pairs as a function of

30   ANALOG VLSI INTEGRATION OF MASSIVE PARALLEL SYSTEMS

the distance. For every size and every distance one sample is obtained per test-chip. Then the standard deviation of the mismatch parameters $\sigma(\Delta V_{T0})$ and $\left(\frac{\sigma(\Delta \beta)}{\beta}\right)$ is calculated by combining the samples of all test-chips and calculating the sample variance from the MAD (median of the absolute differences) [Rey 83] to eliminate the effect of outliers.

At this point it is important to discuss the accuracy of the extracted standard deviations. The estimation of the standard deviation is an application of the estimation of a parameter of the distribution of a random variable [Pap 91]. For a normally distributed random variable, the sample variance s^2 can be used as an estimation for the variance σ^2 and the s^2/σ^2 ratio follows a Chi-squared distribution. This allows to determine the confidence interval for the extracted value for a given confidence level (see also [Per 95]). In our extractions we have aimed at a ±20% accuracy with a 99.7% confidence, which requires a sample size of over 100 samples. For the same confidence level, over 500 samples or test-chips would be required to attain an accuracy of ±10%. These numbers clearly illustrate, the very high measurement effort that has to be done to obtain good quantitative mismatch parameters.

For small transistors the distance effect is completely masked by the large variance due to the small gate area as can be noted in (2.1) and (2.3) and from table 2.2. In the 1.2 μm CMOS technology, for instance, a significant distance dependence is only observed for a 20/20 $\mu m/\mu m$ nMOS transistor as is shown in figure 2.2. A straight line is fitted to the standard deviation data points and the $S_{V_{T0}}$ and S_β from (2.1) and (2.3) are extracted. In table 2.2 the distance dependence model parameters for several technologies are summarized.

E   Mismatch Dependence on Size.   The rows contain transistors of different sizes so that the size dependence of the parameter mismatch can be investigated. For each transistor size, the standard deviation of the parameter mismatch is again estimated from the sample obtained by combining the results of the devices at minimum distance over all test-chips. The same statistical techniques are used as for the distance dependence.

Threshold voltage V_{T0} mismatch.   In figure 2.3 the standard deviation of the threshold voltage mismatch is plotted versus the square root of the effective area for the 10 transistor sizes on the C12 test-chip. The model of (2.1) predicts a linear relation between the $\sigma(V_{T0})$ and the square root of the effective area. For the large transistors, which do not have a minimal width nor a minimal length, the experimental results confirm the model; this also agrees with the experimental results from other authors [Pel 89, Laks86, Mic

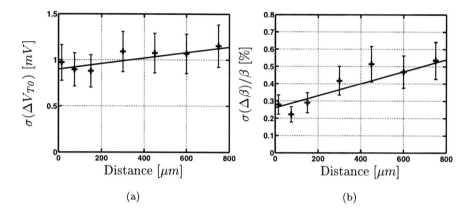

Figure 2.2. Threshold voltage mismatch (a) and current factor mismatch (b) for the 20/20 nMOS transistor versus distance in the 1.2 μm CMOS technology; a straight line is fitted through the measurement points to extract the mismatch distance dependence.

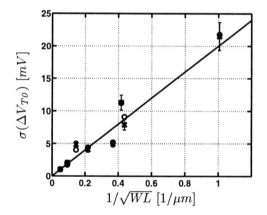

Figure 2.3. The standard deviation of the threshold voltage V_{T0} mismatch versus the square root of the gate area for nMOS transistors in the 1.2 μm CMOS technology; the experimental results are represented by + and the values predicted by the model in equation (2.10) are represented by o; the straight line represents the predictions by the linear model of equation (2.1).

92]. A straight line is fitted and the size proportionality constant is obtained (see table 2.1).

Accurate modeling of V_{T0} mismatch in sub-micron technologies. Narrow channel transistors with a minimal width ($W = 1.4\mu m$, $L = 6.2\mu m$ and $1/\sqrt{W_{eff}L_{eff}} = 0.37/\mu m$) show less mismatch than predicted by the linear model (2.1) as can be verified from the experimental results in figure 2.3; short channel transistors with a minimal gate length ($L = 1.2\mu m$, $W = 6.2\mu m$ and $1/\sqrt{W_{eff}L_{eff}} = 0.42/\mu m$; or $L = 1.2\mu m$, $W = 50\mu m$ and $1/\sqrt{W_{eff}L_{eff}} = 0.15/\mu m$) on the other hand show a significant higher mismatch than predicted by the linear model. In high speed analog designs, the designer prefers to use small gate-lengths so that the highest intrinsic speed f_T for the transistor is obtained [Lak 94]; accurate models for minimum sized transistors are thus necessary.

For the accurate modeling of the threshold mismatch in sub-micron technologies the simple linear model has to be extended for short and narrow channel effects. The threshold voltage is dependent on the flat-band voltage, the surface potential, the depletion charge and the gate capacitance [Lak 94]. It has been verified experimentally that the mismatch of the threshold voltage is mainly attributed on the mismatch of the bulk depletion charges in the two devices [Pel 89, Miz 94, Laks86]. Due to the random process of the ion implantation and the drive-in diffusion process, the doping ions are distributed randomly and the depletion charge fluctuates randomly. The depletion charge follows a Poisson distribution: the mean of the depletion charge is proportional to the gate-area and the bulk doping level; the variance is equal to the square root of the mean of the depletion charge. This leads to the linear model (2.1) where the standard deviation of the threshold voltage is inversely proportional to the gate area.

In sub-micron technologies two effects introduce errors in the model. Due to the presence of the source and the drain diffusion areas and the charge sharing effect, part of the channel depletion charge is not controlled by the gate voltage anymore. For devices with a small gate-length, this charge is a relatively large part of the depletion charge. The depletion charge controlled by the gate is smaller and as a result, the threshold voltage lowers for small gate lengths whereas the variance of the threshold voltage or the V_T mismatch increases.

The depletion charge is not limited to the gate area but due to the fringing field some of the dopant atoms on the side are also depleted. For large widths, the part of the depletion region on the sides is a small percentage of the total depletion region volume. But for narrow channel devices, the side parts are a large percentage of the depletion charge. The depletion charge controlled by the gate is now larger so that the threshold voltage increases and the V_T mismatch decreases for narrow channel devices.

The narrow and short channel effects explain the deviations of the V_T mismatch from the linear model in the experimental data very well. When these

Table 2.3. Mismatch fitting constants for the extended model of equation (2.10) for a 1.2 μm CMOS technology.

PARAMETER	nMOS	pMOS	
A_{1VT}	20	23	$mV\ \mu m$
A_{2VT}	19	20	$mV\ \mu m^{3/2}$
A_{3VT}	18	12	$mV\ \mu m^{3/2}$

effects are modeled quantitatively, the following extended model for the V_{T0} mismatch is obtained [Stey94] [Bas 95]:

$$\sigma^2(\Delta V_{T0}) = \frac{A_{1VT}}{WL} + \frac{A_{2VT}}{WL^2} - \frac{A_{3VT}}{W^2L} \qquad (2.10)$$

In this extended model the second term models the short channel effect for transistors with small gate-lengths; the last term results in a lower mismatch for small gate-widths and models the narrow channel effect. This new model is able to predict the mismatch data within the confidence limits. The model parameters for the 1.2 μm technology are summarized in table 2.3 [Bas 95] and in figure 2.3 the results for the fitting of the new model are represented by the circle; a very good agreement with the experimental data is obtained.

Current factor β mismatch. In figure 2.4 the standard deviation of the current factor mismatch is plotted as function of the square root of the effective area for nMOS devices in the C12 technology. The experimental data fits well to the linear model of (2.3) and no significant deviation for short or narrow channel devices is observed.

F Extraction Validation. The correlation factor between the V_T and β mismatch also has to be computed. In all characterized technologies the correlation factor remains very low and the correlation can be neglected [Bas 95]. This agrees well with the experimental results of other authors [Pel 89, Laks86]. To check the accuracy of the characterization procedure, the measured and the predicted current mismatch are compared. In saturation the predicted variance of the current mismatch is given by (2.29) and (2.31), when the correlation between the parameters is negligible. The standard deviation of the measured current mismatch, for each bias point and for each device size, has been calculated and is compared with the predicted value. In figure 2.5 the measured and predicted standard deviation of the current mismatch is plotted for the

34   ANALOG VLSI INTEGRATION OF MASSIVE PARALLEL SYSTEMS

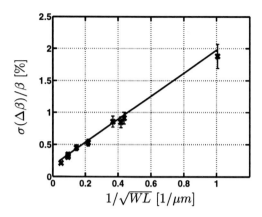

Figure 2.4. The standard deviation of the current factor mismatch as function of the square root of the gate-ares for nMOS devices in the 1.2 μm CMOS technology.

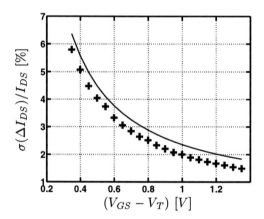

Figure 2.5. Standard deviation of the relative current mismatch versus the gate-overdrive voltage for a 6.2/1.2 nMOS transistor in a 1.2 μm CMOS technology; the crosses indicate the measurements and the solid lines are the theoretical predictions from the extracted mismatch in the parameters.

6.2/1.2 $\mu m/\mu m$ nMOS transistor in the 1.2 μm technology is plotted. The current mismatch is predicted within 20% over the complete measurement range which is within the accuracy limits of the experimental data.

Summary. We conclude that the characterization of transistor mismatch is a tedious process which requires a very large measurement effort. The design, realization and validation of the measurement set-up, the acquisition of the experimental data and the statistical processing of the data have to be performed with great care to avoid errors and systematic effects. Moreover, when migrating towards sub-micron and deep-sub-micron technologies, the standard mismatch models have to checked for their validity; if necessary the effects of the short or narrow channel effects have to be accounted for in model extensions.

On the other hand, in the rest of this chapter, we will demonstrate the importance of a good knowledge of the matching behavior of devices. A circuit designer can only improve the accuracy of circuits by increasing the device area of the devices. Unfortunately, the capacitive load in circuits is also proportional to the area. Generating a signal excursion across a capacitor, results in loading and unloading currents so that the circuit dissipates energy proportional to the capacitive load or the accuracy. The quality of the device matching thus is the ultimate limit for the power consumption of circuits; the impact of transistor mismatch is even more important in high speed applications than thermal noise.

The better the mismatch of devices is modeled and characterized, the smaller area's the designer can safely use while keeping a high circuit yield; consequently the circuits will consume less power for the specified accuracy and speed. When no mismatch data is available, very conservative designs using large devices have to be used and very poor overall circuit performance is achieved.

2.3 IMPLICATIONS OF MISMATCH ON TRANSISTOR BEHAVIOR

In this section the consequence of parameter mismatch on the transistor behavior is calculated for the different possible operation regions of the transistor. First, we introduce the necessary mathematical techniques for mismatch calculations.

2.3.1 Mathematical techniques

Transistor mismatch results in small deviations of the transistor parameters from their nominal value and these deviations result in small deviations of the circuit characteristics from their nominal values. In order to calculate the effect of transistor mismatch on circuit characteristics the following relations are used.

For a circuit characteristic Z, which is defined as $Z = f(x,y)$, the deviation δ_Z in Z due to a deviation δ_x in x and δ_y in y is calculated as:

$$\delta_Z = \frac{\partial f}{\partial x}\delta_x + \frac{\partial f}{\partial y}\delta_y \qquad (2.11)$$

For independent and normally distributed deviations in x and y the standard deviation of Z is:

$$\sigma^2(Z) = \left(\frac{\partial f}{\partial x}\right)^2 \sigma^2(x) + \left(\frac{\partial f}{\partial y}\right)^2 \sigma^2(y) \qquad (2.12)$$

For two characteristics Z_1 and Z_2 defined as $Z_1 = f(x_1, y_1)$ and $Z_2 = f(x_2, y_2)$ the following relations are derived applying (2.11) and (2.12), with $\Delta x = x_1 - x_2$ and $\Delta y = y_1 - y_2$:

$$\Delta Z = Z_1 - Z_2 = \frac{\partial f}{\partial x}\Delta x + \frac{\partial f}{\partial y}\Delta y \qquad (2.13)$$

$$\sigma^2(\Delta Z) = \left(\frac{\partial f}{\partial x}\right)^2 \sigma^2(\Delta x) + \left(\frac{\partial f}{\partial y}\right)^2 \sigma^2(\Delta y) \qquad (2.14)$$

$$\sigma(\Delta Z) = \sqrt{2}\sigma(Z) \qquad (2.15)$$

These relations allow us to calculate the effect of parameter mismatches and transistor mismatches on the transistor or circuit behavior.

2.3.2 Implications on transistor behavior

A circuit designer can bias a transistor in two ways: for current biasing the current through the device is imposed and the terminal voltages are the dependent variables; for voltage biasing the terminal voltages are imposed and the current is the dependent variable. For the sake of simplicity of the equations and calculations, the source and bulk are supposed connected so that $V_{SB} = 0$ and no bulk-effect occurs. However, if a bulk-effect does occur in the circuit, the extra mismatch due to the mismatch in the bulk-effect coefficients can in first order simply be considered as an extra degradation of the V_T matching of the transistors, so that most equations can still be used. Of course, for the optimization of the biasing, the dependence of the bulk-effect on the bias voltages should be taken into account and slightly different results will be obtained.

For voltage biasing the terminal voltages are imposed and the current I_{DS} is the dependent variable as is illustrated in figure 2.6(a); the current depends on the terminal voltages and on the transistor parameters:

$$I_{DS} = f(V_{GS}, V_{DS}, V_{T0}, \beta) \qquad (2.16)$$

For a pair of transistors with an identical V_{GS}, the difference in their currents is calculated using (2.13):

$$\Delta I_{DS} = \frac{\partial I_{DS}}{\partial \beta}\Delta\beta + \frac{\partial I_{DS}}{\partial V_{T0}}\Delta V_{T0} \qquad (2.17)$$

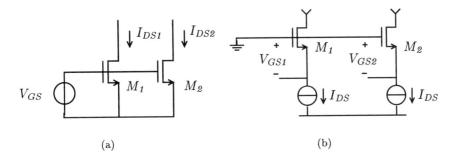

Figure 2.6. (a) Two transistors biased with equal gate-source voltage; (b) two transistors biased with equal drain-source current.

From the device equations in appendix A one can conclude for all regions of operation that:

$$\frac{\partial I_{DS}}{\partial V_{T0}} = \frac{\partial I_{DS}}{\partial V_{GS}} = -g_m \qquad (2.18)$$

$$\frac{\partial I_{DS}}{\partial \beta} = \frac{I_{DS}}{\beta} \qquad (2.19)$$

so that the relative current difference in all operating regions of the transistor is expressed by:

$$\left(\frac{\Delta I_{DS}}{I_{DS}}\right) = \left(\frac{\Delta \beta}{\beta}\right) - \left(\frac{g_m}{I_{DS}}\right) \Delta V_{T0} \qquad (2.20)$$

and the variance of the current difference is:

$$\left(\frac{\sigma(\Delta I_{DS})}{I_{DS}}\right)^2 = \left(\frac{\sigma(\Delta \beta)}{\beta}\right)^2 + \left(\frac{g_m}{I_{DS}}\right)^2 \sigma^2(\Delta V_{T0}) \qquad (2.21)$$

In a current biasing scheme as illustrated in figure 2.6(b) the current is imposed and the gate-source voltage V_{GS} is the dependent variable. Expressions for the calculation of the V_{GS} as a function of the current are not always available. However (2.16) can be rewritten as an implicit function for V_{GS}:

$$I_{DS} - f(V_{GS}, V_{DS}, V_{T0}, \beta) = 0 \qquad (2.22)$$

and the partial derivatives of V_{GS} can be expressed as:

$$\frac{\partial V_{GS}}{\partial \beta} = -\left(\frac{\partial I_{DS}}{\partial \beta}\right)\left(\frac{\partial I_{DS}}{\partial V_{GS}}\right)^{(-1)} \qquad (2.23)$$

$$= \frac{-I_{DS}}{\beta g_m} \qquad (2.24)$$

$$\frac{\partial V_{GS}}{\partial V_{T0}} = -\left(\frac{\partial I_{DS}}{\partial V_{T0}}\right)\left(\frac{\partial I_{DS}}{\partial V_{GS}}\right)^{(-1)} \qquad (2.25)$$

$$= 1 \qquad (2.26)$$

so that the gate-source voltage difference and its variance become:

$$\Delta V_{GS} = \Delta V_{T0} - \left(\frac{I_{DS}}{g_m}\right)\left(\frac{\Delta \beta}{\beta}\right) \qquad (2.27)$$

$$\sigma^2(\Delta V_{GS}) = \sigma^2(\Delta V_{T0}) + \left(\frac{I_{DS}}{g_m}\right)^2 \left(\frac{\sigma(\Delta \beta)}{\beta}\right)^2 \qquad (2.28)$$

2.3.2.1 Strong inversion. For a transistor biased in strong inversion and in saturation, the g_m/I_{DS} is $2/(V_{GS} - V_T)$ (see (A.22)) so that (2.21) and (2.28) can be rewritten as:

$$\left(\frac{\sigma(\Delta I_{DS})}{I_{DS}}\right)^2 = \left(\frac{\sigma(\beta)}{\beta}\right)^2 + \frac{4\sigma^2(V_{T0})}{(V_{GS} - V_T)^2} \qquad (2.29)$$

$$\sigma^2(\Delta V_{GS}) = \sigma^2(V_{T0}) + \frac{(V_{GS} - V_T)^2}{4}\left(\frac{\sigma(\beta)}{\beta}\right)^2 \qquad (2.30)$$

Substituting the models of (2.1) and (2.3) in (2.29) and (2.30), we obtain for closely spaced devices:

$$\left(\frac{\sigma(\Delta I_{DS})}{I_{DS}}\right)^2 = \frac{1}{WL}\left(A_\beta^2 + \frac{4A_{VT0}^2}{(V_{GS} - V_T)^2}\right) \qquad (2.31)$$

$$\sigma^2(\Delta V_{GS}) = \frac{1}{WL}\left(A_{VT0}^2 + \frac{(V_{GS} - V_T)^2 A_\beta^2}{4}\right) \qquad (2.32)$$

The accuracy of the gate voltage or drain current is dependent on the bias point of the devices or $(V_{GS} - V_T)$ and on their gate-area. For a technology a *corner* gate-drive voltage $(V_{GS} - V_T)_m$ is defined for which the effect of the V_{T0} and β mismatch on the gate voltage or drain current is of equal size:

$$(V_{GS} - V_T)_m = 2A_{VT0}/A_\beta \qquad (2.33)$$

In (2.31) and (2.32) we observe that for a circuit with a bias point with a $(V_{GS} - V_T)$ smaller than $(V_{GS} - V_T)_m$ the effect of the V_{T0} mismatch is dominant, whereas for a $(V_{GS} - V_T)$ larger than $(V_{GS} - V_T)_m$ the effect of the β mismatch dominates. In table 2.1 the values of $(V_{GS} - V_T)_m$ are listed for a few CMOS technologies. It is clear that in practical circuits the $(V_{GS} - V_T)$ will be smaller than $(V_{GS} - V_T)_m$ so that the V_{T0} mismatch is dominant over the β mismatch for the calculation of the accuracy of the circuit behavior. In practice, equations (2.31) and (2.32) can be approximated by:

$$\left(\frac{\sigma(\Delta I_{DS})}{I_{DS}}\right)^2 \approx \frac{4A_{VT0}^2}{WL(V_{GS} - V_T)^2} \quad (2.34)$$

$$\sigma^2(\Delta V_{GS}) \approx \frac{A_{VT0}^2}{WL} \quad (2.35)$$

The approximation error due to neglecting the β mismatch on the standard deviation σ for the above equations is equal to $((V_{GS} - V_T)/(V_{GS} - V_T)_m)^2/2$, and is small for typical transistor bias conditions; for a nMOS transistor in the 0.7 μm technology biased with a $(V_{GS} - V_T)$ of 0.2 V the error on σ is only 1 %.

2.3.2.2 Weak inversion. When the transistors are biased in weak inversion the g_m/I_{DS} is $1/(nU_T)$ (see (A.4)) so that (2.21) and (2.28) can be rewritten as:

$$\left(\frac{\sigma(\Delta I_{DS})}{I_{DS}}\right)^2 = \left(\frac{\sigma(\beta)}{\beta}\right)^2 + \frac{\sigma^2(V_{T0})}{(nU_T)^2} \quad (2.36)$$

$$\sigma^2(\Delta V_{GS}) = \sigma^2(V_{T0}) + (nU_T)^2 \left(\frac{\sigma(\beta)}{\beta}\right)^2 \quad (2.37)$$

By substituting (2.1) and (2.3) in (2.36) and (2.37), the relative current variation and gate voltage variation of two closely spaced devices become:

$$\left(\frac{\sigma(\Delta I_{DS})}{I_{DS}}\right)^2 = \frac{1}{WL}\left(A_\beta^2 + \frac{A_{VT0}^2}{(nU_T)^2}\right) \quad (2.38)$$

$$\sigma^2(\Delta V_{GS}) = \frac{1}{WL}(A_{VT0}^2 + (nU_T)^2 A_\beta^2) \quad (2.39)$$

The weak-inversion slope parameter n typically has values from 1 to 2 and U_T is 25.8 mV at room temperature. From table 2.1 we can conclude that the V_{T0} mismatch dominates the accuracy calculations so that (2.38) and (2.39) can be

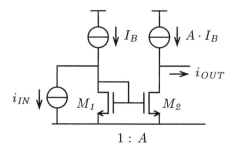

Figure 2.7. Basic current amplifier.

simplified to:

$$\left(\frac{\sigma(\Delta I_{DS})}{I_{DS}}\right)^2 \approx \frac{A_{VT0}^2}{WL(nU_T)^2} \tag{2.40}$$

$$\sigma^2(\Delta V_{GS}) \approx \frac{A_{VT0}^2}{WL} \tag{2.41}$$

The approximation error due to neglecting the β mismatch on the standard deviation σ for the above equations is equal to $(2nU_T/(V_{GS} - V_T)_m)^2/2$; for a nMOS transistor in the 0.7 μm technology the error on the σ is only 0.1 % at room temperature.

2.4 IMPLICATIONS OF TRANSISTOR MISMATCH ON THE BEHAVIOR AND DESIGN OF ELEMENTARY STAGES

In the forthcoming sections the implications of transistor mismatch on the speed, power consumption and accuracy of elementary current and voltage processing stages is studied. This allows to draw guidelines for optimal design of these circuits. Furthermore it provides the background for a discussion of the implications of transistor mismatch on the performance of general analog VLSI systems [Kin 96d].

2.4.1 Current processing circuits

The current amplifier, represented in figure 2.7, is a basic current processing stage. The output transistor M_2 is a parallel connection of A unit-transistors of the same size as M_1, with a gate-width W and gate-length L, so that the current amplification factor is A.

The speed performance of the current amplifier is the highest frequency that can be processed by the current amplifier, and depends on the bandwidth.

The bandwidth of this circuit is in first-order determined by the g_m of the input transistor M_1 and the parallel connection of the gate capacitors of both transistors. For transistors biased in strong inversion, the bandwidth of the amplifier is:

$$\begin{aligned} \text{BW} &= \frac{g_{m1}}{2\pi(C_{GS1} + C_{GS2})} \\ &= \frac{3I_B}{2\pi(A+1)(V_{GS} - V_T)C_{ox}WL} \end{aligned} \quad (2.42)$$

The DC power consumption of the amplifier is :

$$P = (A+1)I_B V_{DD} \quad (2.43)$$

The relative accuracy of the current processing is determined by the maximal input signal RMS value I_{inRMS} and the 3σ value of the input referred offset current I_{OS} and is defined as:

$$\text{Acc}_{\text{rel}} = \frac{I_{inRMS}}{3\sigma(I_{OS})} \quad (2.44)$$

By using the 3σ value of the offset current, the accuracy specification is met with a probability of about 99.7%. This probability that a circuit block meets its specifications, has a direct impact on the yield of the total chip or system. In complex systems with many stages, an even higher probability can be necessary to obtain a high yield and more than the 3σ has to be accounted for in (2.44).

Due to the effect of mismatches in the transistors, an error occurs in the current mirroring and for a zero input current a non-zero output current exists. The input referred offset current I_{OS} is by definition the current that has to be applied to the input to obtain a zero output current; it has to compensate for the variation in the current of M_1 and M_2. To calculate the I_{OS}, we first calculate the errors in the currents of M_1 and M_2. The standard deviation of the current in transistor M_1 and in a unit-transistor of M_2 is derived from (2.34):

$$\sigma(I_{UNIT}) = \frac{1}{\sqrt{2}} I_B \frac{2A_{VT0}}{(V_{GS} - V_T)\sqrt{WL}} \quad (2.45)$$

where the term $1/\sqrt{2}$ is necessary to calculate the variance of the parameter from the variance of the difference of parameters – see (2.15). The total current in M_2 is the sum of the currents of the individual unit transistors, which are statistically independent quantities, so that the standard deviation of the current of M_2 is $\sigma(I_{OUT}) = \sqrt{A}\sigma(I_{UNIT})$. Since the current amplification is A, we can express the standard deviation of the input offset current as follows:

$$\sigma(I_{OS}) = \sqrt{\frac{\sigma^2(I_{OUT})}{A^2} + \sigma^2(I_{UNIT})}$$
$$= I_B \frac{\sqrt{2}A_{VT0}}{(V_{GS} - V_T)\sqrt{WL}}\sqrt{\frac{A+1}{A}} \quad (2.46)$$

For a typical bias modulation index of 1/2 the I_{inRMS} is $I_B/(2\sqrt{2})$ and the relative accuracy (Acc_{rel}) of the current processing then becomes:

$$\text{Acc}_{\text{rel}} = \frac{\sqrt{WL}(V_{GS} - V_T)}{12A_{VT0}}\sqrt{\frac{A+1}{A}} \quad (2.47)$$

With the expressions for the different circuit performance parameters at hand, we can now develop a relation for the total performance or quality of the circuit design. The quality of the circuit is better if its bandwidth, gain and accuracy are large and its power consumption is low. When we consider the ratio *Speed·Accuracy²/Power* for the current amplifier, we obtain:

$$\frac{\text{BW}\text{Acc}_{\text{rel}}^2}{P} = \frac{(V_{GS} - V_T)}{96\pi C_{ox}A_{VT0}^2 V_{DD}}\frac{A}{(A+1)^3} \quad (2.48)$$

which we can rewrite for large gains A as:

$$\frac{\text{Gain}^2\text{BW}\text{Acc}_{\text{rel}}^2}{P} = \frac{(V_{GS} - V_T)}{96\pi C_{ox}A_{VT0}^2 V_{DD}} \quad (2.49)$$

Very important conclusions for the design of a current amplification stage can be drawn from this relation:

- The *total* performance of the amplifier is only dependent on technology constants and on the chosen bias point of the stage and is independent of the transistor sizes. To obtain the best *total* performance, a current processing stage must be designed with a *large* $(V_{GS} - V_T)$. It is common knowledge that to improve the accuracy of a current mirror a large $(V_{GS} - V_T)$ has to be used [Lak 94]. From (2.29) one can indeed conclude that the accuracy performance of a current mirror is improved by increasing $(V_{GS} - V_T)$. On the other hand, increasing $(V_{GS} - V_T)$ reduces the g_m of the stage so that the speed performance is degraded and that a higher bias current must be used to obtain a certain speed, resulting in higher power consumption. However, (2.49) shows that increasing $(V_{GS} - V_T)$ will result in the best possible trade-off between speed, accuracy, gain and power consumption for a current processing stage !

For a current processing stage the $(V_{GS} - V_T)$ is typically limited to $V_{DD}/2^\dagger$. As a result, the optimal performance becomes:

$$\frac{\text{Gain}^2 \text{BW} \text{Acc}_{\text{rel}}^2}{P} = \frac{1}{192\pi C_{ox} A_{VT0}^2} \qquad (2.50)$$

- The different performance specifications of a current amplifier are linked and depend on technological and physical constants only ! Equation (2.50) shows that a designer can only trade one specification for the other. Due to the impact of transistor mismatch, he cannot choose the different performance specifications independently ! When we rewrite (2.50) as:

$$P = 192\pi C_{ox} A_{VT0}^2 \text{Gain}^2 \text{BW} \text{Acc}_{\text{rel}}^2 \qquad (2.51)$$

Transistor mismatch puts a boundary on the minimal power consumption by a current amplification stage for a given gain, speed and accuracy specification.

- An important limitation of current processing stages is the quadratic dependence of the power consumption on the gain of the amplifier in (2.51). The fundamental reason for the appearance of this term is related to the physics of the MOS transistor. By changing the gate voltage, the conductivity of the channel is controlled and thus the current is controlled. On the physical level, a MOS transistor acts as a voltage dependent current source or a transconductor. If we try use this device as a current amplifier, we basically operate the device in an unnatural way. The only way to make current gain is by a parallel connection of several transistors to the gate of a diode-connected transistor, that does the current to voltage conversion. The more gain we try to make, the more load we put on this gate for the same transconductance and the speed in (2.42) reduces. Moreover, the more gain, the larger the bias current in the output transistor and the larger the power consumption in (2.43).

The accuracy in (2.47) on the other hand, is in first order independent of the realized gain. The input signal swing is only limited by the modulation index we can allow. The maximal modulation index depends on the distortion as is discussed in section 2.7. The bias current of the output stage scales with the gain so that the modulation index at the output is the same as at the

---

†for small feature sizes, the appearance of velocity saturation [Lak 94] can further reduce the maximal $(V_{GS} - V_T)$ that can be used under which the above derivations remain valid.

input. As a result the gain does not influence the maximal signal swing or the relative accuracy.

Due to the effect of the gain on the bandwidth and the power consumption, a quadratic dependence on the gain of the quality of the circuit in (2.49) and of the the minimal power consumption of the circuit in (2.51) emerges.

Exact case When the full expression for the variance of the relative current difference (2.31) is used, including both the V_{T0} and β mismatch, the following expression for the quality of a current amplifier with a large gain is derived:

$$\frac{\text{Gain}^2 \text{BW Acc}_{\text{rel}}^2}{P} = \frac{1}{24\pi C_{ox} V_{DD} \left(A_\beta^2 (V_{GS} - V_T) + \frac{4 A_{VT0}^2}{(V_{GS} - V_T)} \right)} \quad (2.52)$$

The value of $(V_{GS} - V_T)$ can now be optimized to obtain a maximal total circuit performance: starting with a small $(V_{GS} - V_T)$ and increasing it, the term proportional to A_{VT0} in the denominator of (2.52) decreases and the quality improves; for large values of $(V_{GS} - V_T)$, however, the first term proportional to A_β increases and the quality decreases again. An optimal value for $(V_{GS} - V_T)$ exists and can easily be calculated: the optimum in circuit performance is reached for a $(V_{GS} - V_T)$ equal to $(V_{GS} - V_T)_m$ - see (2.33), and we then obtain the best possible total circuit performance:

$$\frac{\text{Gain}^2 \text{BW Acc}_{\text{rel}}^2}{P} = \frac{1}{96\pi C_{ox} V_{DD} A_\beta A_{VT0}} \quad (2.53)$$

In the previous section, taking only the effect of V_T offset into account, we concluded that for current processing circuits the $(V_{GS} - V_T)$ of the stage has to be maximized to optimize performance. However, by taking also the effect of β mismatch into account, we conclude that $(V_{GS} - V_T)$ should not be increased beyond $(V_{GS} - V_T)_m$. Furthermore, if the supply voltage allows it, a current processing stage should be biased with $(V_{GS} - V_T) = (V_{GS} - V_T)_m$ which yields the optimum in the trade-off between the different specifications.

2.4.2 Voltage processing circuits

The inverting voltage amplifier stage of figure 2.8 is a basic voltage processing stage. It consists of an operational amplifier (opamp) with a negative resistive feedback. The input voltage is converted into a current by R_1 and the virtual ground that is created at the negative input terminal of the opamp. This current is converted into the output voltage by R_2 so that the closed loop amplification of the system A_{CL} is determined by the ratio R_2/R_1, which is

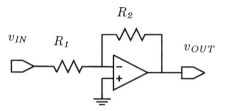

Figure 2.8. System schematic for a basic voltage amplification block.

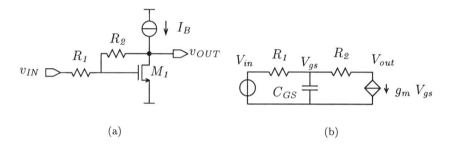

Figure 2.9. (a) One transistor voltage amplifier implementation; (b) small signal equivalent.

well controlled over process variations. This behavior of the system is observed as long as the open-loop gain of the opamp is much larger than the closed loop gain.

2.4.2.1 One transistor voltage amplifier. The most simple implementation for the operational amplifier in figure 2.8 is obtained with a single transistor as in figure 2.9(a). From the small signal equivalent in figure 2.9(b), we derive the frequency response of the system:

$$A(s) = \frac{V_{out}}{V_{in}}$$
$$= -\frac{R_2}{R_1}\left(\frac{1 - 1/(g_m R_2)}{1 + 1/(g_m R_1) + s/(g_m/C_{GS})}\right) \quad (2.54)$$

We observe in (2.54) that in order to define the amplification accurately, the $1/g_m$ of the transistor has to be chosen considerably smaller than the resistors R_1 and R_2. The low-frequency closed-loop amplification of the stage is then $A_{CL} = R_2/R_1$ and the closed-loop frequency response is approximately given by:

$$A(s) \approx -\frac{R_2}{R_1}\left(\frac{1}{1+s/(g_m/C_{GS})}\right) \quad (2.55)$$

so that we obtain the following relation for the the bandwidth of the amplifier:

$$\text{BW} \approx g_m/(2\pi C_{GS}) \quad (2.56)$$

A   Strong inversion. First we consider the performance of this voltage amplifier when the transistor is biased in strong inversion. Using the relations for the g_m and C_{GS} in strong inversion from appendix A, we can express the bandwidth of the amplifier as follows:

$$\text{BW} = \frac{3I_B}{2\pi(V_{GS}-V_T)WLC_{ox}} \quad (2.57)$$

The DC power consumption of the circuit is:

$$P = V_{DD}I_B \quad (2.58)$$

The relative accuracy (Acc$_{\text{rel}}$) of the circuit is determined by the offset voltage of the opamp (V_{OS}) with respect to the maximal RMS value of the input signal (V_{inRMS}). The maximal voltage swing at the output is in first order: $V_{outPP} = V_{DD}$, so that $V_{inRMS} = V_{DD}/(2\sqrt{2}\text{Gain})$. The standard deviation of the offset voltage of the one transistor opamp is determined by equation (2.35), but from (2.15) we conclude that the offset voltage is $\sqrt{2}$ times smaller since the variation on V_{GS} is important. The expression for the relative accuracy can now be evaluated as:

$$\text{Acc}_{\text{rel}} = \frac{V_{inRMS}}{3\sigma(V_{OS})} \quad (2.59)$$

$$= \frac{V_{DD}\sqrt{WL}}{6A_{VT0}\text{Gain}} \quad (2.60)$$

To obtain a better insight in the possible trade-offs between the different circuit parameters, we consider again the total performance of the voltage amplifier. Combining (2.57), (2.58) and (2.60), we obtain:

$$\frac{\text{Gain}^2\text{BW}\,\text{Acc}_{\text{rel}}^2}{P} = \frac{V_{DD}}{24\pi(V_{GS}-V_T)A_{VT0}^2 C_{ox}} \quad (2.61)$$

From (2.61) some very important lessons can be learned for the design of a voltage amplifier:

IMPLICATIONS OF MISMATCH ON ANALOG VLSI 47

- The *total* performance of a voltage amplifier is maximized by *lowering* $(V_{GS} - V_T)$. As far as speed requirements allow it the stage should be biased with the lowest possible $(V_{GS} - V_T)$. Again it is common knowledge that to achieve a low offset voltage a small $(V_{GS} - V_T)$ has to be used as can also be derived from (2.32) [Lak 94]. However, (2.61) shows that not only a good accuracy but the best trade-off between speed, gain, accuracy and power consumption is obtained for a small $(V_{GS} - V_T)$. A typical minimal value of $(V_{GS} - V_T) = 0.2\ V$ is derived from device physics - see Appendix A. The best attainable performance in strong inversion is:

$$\frac{\text{Gain}^2 \text{BW} \text{Acc}_{\text{rel}}^2}{P} = \frac{5V_{DD}}{24\pi A_{VT0}^2 C_{ox}} \tag{2.62}$$

- Also for voltage designs the different performance specifications are linked by physical and technological constants only ! The designer can - once the optimal bias is chosen - not optimize the different specifications independently but can only trade one specification for an other. When we rewrite (2.62), we obtain that the minimal required power consumption is fixed for a given gain, speed and accuracy by the impact of transistor mismatch:

$$P = \frac{24\pi}{5V_{DD}} A_{VT0}^2 C_{ox} \text{Gain}^2 \text{BW} \text{Acc}_{\text{rel}}^2 \tag{2.63}$$

- In (2.63) we observe again a quadratic dependence of the power consumption on the gain of the voltage amplifier. In a voltage amplifier nor the power consumption, nor the bandwidth are in first order dependent on the realized gain. However, when the gain is increased, the maximal input signal reduces proportionally since the maximal output swing is limited by the supply voltage. This limits the maximal input signal swing; consequently, the maximal attainable accuracy for a given power supply voltage reduces proportional to the gain (2.60) and the power consumption increases quadratically with the gain (2.63).

B Weak inversion. In the previous section it was shown that the *total performance* of a voltage amplifier improves for smaller $(V_{GS} - V_T)$. Consequently, the best total performance is obtained when the transistor is biased in weak-inversion.

The relations for the power consumption (2.58) and for the relative accuracy (2.60) remain valid. Since source and bulk are connected, the capacitance that determines the speed performance in weak inversion is the gate-bulk capacitance, which is:

$$C_{GB} = \frac{(n-1)}{n} C_{ox} WL \tag{2.64}$$

48 ANALOG VLSI INTEGRATION OF MASSIVE PARALLEL SYSTEMS

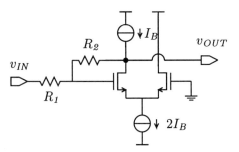

Figure 2.10. Basic voltage amplification stage implemented with a differential pair amplifier.

With the expression for g_m in weak inversion, we derive the following relation for the bandwidth of the amplifier operating in weak-inversion:

$$\text{BW} = \frac{g_m}{2\pi C_{GB}} = \frac{I_B}{2\pi U_T (n-1) C_{ox} WL} \qquad (2.65)$$

And we can now calculate the total performance of the amplifier in weak inversion as:

$$\frac{\text{Gain}^2 \text{BW} \text{Acc}_{\text{rel}}^2}{P} = \frac{V_{DD}}{72\pi (n-1) U_T A_{VT0}^2 C_{ox}} \qquad (2.66)$$

For the 0.7 μm technology e.g. the technology constants are $V_{DD} = 5\ V$, $C_{ox} = 2\ fF/\mu m^2$, $A_{VT0} = 13\ mV\mu m$, and n is approximately 1.5 and $U_T = 25.8\ mV$ at room temperature; the optimal quality in weak inversion calculated with (2.66) is then 5 times better than the best quality in strong inversion calculated with (2.62). This implies that for the same gain, bandwidth and accuracy specifications 5 times less power is required for an amplifier operated in weak inversion.

2.4.2.2 Differential pair voltage amplifier. A simple differential implementation of the operational amplifier in figure 2.8 is a differential pair as is shown in figure 2.10. Similarly as for the one transistor amplifier the $1/g_m$ of the input transistors is designed larger than the value of the resistors resistors R_1 and R_2 to obtain a well controlled closed-loop gain. As a result the bandwidth of the amplifier in a high precision design is limited by the pole caused by the input capacitance and it is easily calculated that: $\text{BW} = g_m/(2\pi C_{GS})$.

The power consumption of the circuit is $P = 2V_{DD}I$. The relative accuracy (Acc_{rel}) of the circuit is determined by the offset voltage of the opamp, or the

V_{T0} mismatch of the input transistors, in respect to the maximal RMS of the input signal, which is in first order $V_{DD}/(2\sqrt{2}\ \text{Gain})$:

$$\text{Acc}_{\text{rel}} = \frac{V_{DD}}{6\sqrt{2}\ \sigma(V_{OS})\ \text{Gain}} \\ = \frac{V_{DD}\sqrt{WL}}{6\sqrt{2}\ A_{VT0}\ \text{Gain}} \quad (2.67)$$

For transistors biased in strong inversion and saturation, the quality of the design is again expressed as:

$$\frac{\text{Gain}^2 \text{BW}\,\text{Acc}_{\text{rel}}^2}{P} = \frac{V_{DD}}{96\pi(V_{GS}-V_T)A_{VT0}^2 C_{ox}} \quad (2.68)$$

The *total* performance of this voltage processing stage is again maximized by *lowering* $(V_{GS}-V_T)$ towards operation in weak inversion. This result shows that the differential implementation consumes about 4 times more power for the same gain, bandwidth and accuracy specifications as the single transistor implementation (see equation (2.61)). The power consumption is doubled since two transistors require two times the bias current and their offset voltage is $\sqrt{2}$ as for the single transistor (see equation (2.15)); both effects result in 4 fold increase of the power consumption.

Summary. To obtain a low power consumption, voltage circuits must be operated in weak inversion as much as possible. However, the maximal attainable frequency in weak inversion is limited, since the maximal current in weak inversion is also limited. Weak inversion operation is only possible for relatively low speed applications. In section 2.4.4.2 the limitations on the use of (2.61) and (2.66) to calculate the necessary power consumption are studied in detail. A second practical, but important, limitation of the use of weak inversion, is the limited availability from industrial chip foundries of good model parameters for the simulation of transistors in weak inversion.

Up to now we have discussed the mismatch limitation on the trade-offs between the different performance parameters for simple signal processing stages. This makes the analytical analysis straightforward and closed-form expressions are obtained. We will now proceed with the analysis of a more complex circuit implementation of a load compensated OTA. However, increasing the circuit complexity, increases the degrees of freedom in the design and involves the design of several stages; in the calculations some reasonable assumptions and approximations are made to obtain again closed-form expressions.

2.4.2.3 Load compensated OTA voltage processing stage. The operational amplifier that is used in the basic voltage amplifier of figure 2.8 can

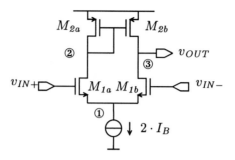

Figure 2.11. Load compensated OTA.

also be implemented with the operational transconductance amplifier represented in figure 2.11. It is the basic schematic for many more sofisticated load compensated operational amplifiers. As such, its design and the trade-off between its specifications, is representative for the design procedure of many multi-stage operational amplifiers. The circuit operates as follows: transistors M_{1a-b} transform the differential component of the input voltages into a differential current; they act as a voltage to current converter. The signal current is converted to a single ended output voltage by the current mirror M_{2a-b} and the output conductance of M_{1b} and M_{2b}.

These amplifiers are typically used in a feedback configuration like e.g. the configuration of figure 2.8. The stability of the feedback system is then first concern of the circuit designer; an amplifier is only useful if its stability is guaranteed. Due to the high complexity of the circuit, its transfer function is of higher order. Basically, a pole is associated with each circuit node. The presence of a second and higher order poles in the open-loop transfer function implies that the open-loop frequency response of the OTA has to be adapted in order to obtain a safe phase and gain margin for all possible feedback configurations. The maximal speed or frequency performance that can be attained is then determined by the frequency of the second pole in the open-loop transfer function. The gain-bandwidth (GBW) of the amplifier must be made K_{stab} times smaller than the second pole and K_{stab} is at least 2 for a phase margin better than 60 degrees.

The second-pole (f_2) in the open-loop transfer function is located at node ②, the gates of transistors M_{2a} and M_{2b}, and is calculated as:

$$f_2 = \frac{g_{m2}}{2\pi(C_{GS2a} + C_{GS2b})}$$
$$= \frac{3I_B}{4\pi C_{ox} W_2 L_2 (V_{GS} - V_T)_2} \tag{2.69}$$

so that the maximal GBW of the amplifier becomes:

$$\text{GBW} = \frac{f_2}{K_{stab}}$$
$$= \frac{3I_B}{4\pi K_{stab} C_{ox} W_2 L_2 (V_{GS} - V_T)_2} \tag{2.70}$$

The accuracy of the amplifier depends on the equivalent input referred offset voltage (V_{OS}). The V_{OS} is determined by the voltage matching of the transistor pair M_{1a-b} and the accuracy of the current mirror M_{2a-b}. For the calculation of V_{OS}, the internal voltage gain of the amplifier A_{in} from the input to gate of M_{2a-b} is an important factor. It determines the attenuation of the influence of the errors in the second stage on the input referred offset. In order to simplify the expressions we assume that all devices have equal length ($L_1 = L_2$), which is a reasonable assumption for high speed designs where all signal transistors are designed with a minimal length to obtain a maximal transconductance and minimal node and input capacitances. All transistors have the same bias current, so that the A_{in} is expressed by:

$$A_{in} = \frac{g_{m1}}{g_{m2}} = \frac{(V_{GS} - V_T)_2}{(V_{GS} - V_T)_1} = \sqrt{\frac{W_1}{W_2}} \tag{2.71}$$

The standard deviation of the offset voltage is calculated as follows:

$$\sigma^2(V_{OS}) = \sigma^2(V_{T01}) + \left(\frac{\sigma(V_{T02})}{A_{in}}\right)^2 \tag{2.72}$$

$$= \frac{A_{VT0n}^2}{W_1 L_1} + \frac{A_{VT0p}^2}{W_2 L_2 A_{in}^2} \tag{2.73}$$

$$= \frac{1}{W_2 L_2 A_{in}^2} \left(A_{VT0n}^2 + A_{VT0p}^2\right) \tag{2.74}$$

Equation (2.72) shows that by increasing the internal gain A_{in}, the effect of the mismatch in the current mirror on the input referred offset is lowered. However, from (2.71) we conclude that the gain can only be increased by decreasing

52   ANALOG VLSI INTEGRATION OF MASSIVE PARALLEL SYSTEMS

the width W_2 and consequently by decreasing the area of the current mirror transistors since the transistors have the same lengths; the increase in gain goes at the cost of a reduction of the matching; therefor the nMOS as well as the pMOS matching, expressed by respectively A_{VT0n} and A_{VT0p}, are equally important in the final expression for the offset voltage (2.74).

Using the expression (2.74), the relative accuracy (Acc_{rel}) is calculated from (2.59). The DC power consumption of the amplifier is given by:

$$P = 2I_B V_{DD} \tag{2.75}$$

The quality of the circuit implementation is evaluated from:

$$\frac{\text{Gain}^2 \text{GBW} \text{Acc}_{\text{rel}}^2}{P} = \frac{V_{DD}}{192\pi K_{stab} C_{ox}(A_{VT0n}^2 + A_{VT0p}^2)} \cdot \left(\frac{A_{in}}{(V_{GS} - V_T)_2}\right) \tag{2.76}$$

where Gain is the gain in the closed loop configuration determined by the applied feedback.

The speed performance of the total voltage processing circuit, as depicted in figure 2.8, with an OTA is determined by the maximal frequency for which the loop gain is larger than 1. Since the GBW of open-loop transfer function of the OTA is fixed, we obtain the following expression for the BW of the total amplifier:

$$\text{BW} = \frac{\text{GBW}}{\text{Gain}} \tag{2.77}$$

Substituting this result in (2.76), we derive the following relation for the voltage processing system:

$$\frac{\text{Gain}^3 \text{BW} \text{Acc}_{\text{rel}}^2}{P} = \frac{V_{DD}}{192\pi K_{stab} C_{ox}(A_{VT0n}^2 + A_{VT0p}^2)} \cdot \left(\frac{A_{in}}{(V_{GS} - V_T)_1}\right) \tag{2.78}$$

This expression leads to the following conclusions:

- The quality of the design improves by biasing the input transistors M_{1a-b}, which operate in *voltage mode*, with a *low* $(V_{GS} - V_T)$.

- By choosing a high internal gain and consequently by biasing the transistors M_{2a-b}, which operate in *current mode*, with a *high* $(V_{GS} - V_T)$ better performance is obtained. The choice of the gain in multi-stage amplifiers is further examined in section 2.5.1.

- The minimal power to obtain a given speed and accuracy depends on the *cubic* of the gain of the voltage processing block. As in the one transistor

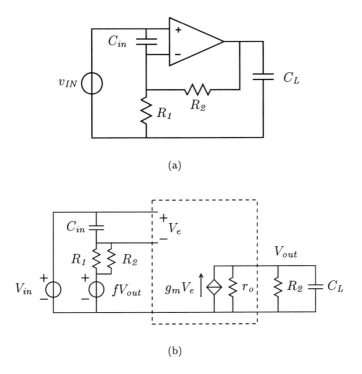

Figure 2.12. (a) OTA based voltage processing system schematic; (b) small signal equivalent where $f = R_1/(R_1 + R_2)$.

implementation a second order dependence is due to the limited output swing of the amplifier, which results in a maximal input signal and relative accuracy inversely proportional to the gain. However an extra gain dependence is due to the presence of a second pole in the amplifier; the power consumption is proportional to the GBW of the open-loop but in closed loop the bandwidth is the gain times smaller. Due to the higher complexity which brings extra constraints relating to the stability into the design, more power is consumed.

2.4.3 Feedback systems and general OTA design

In the previous sections we showed that the *Speed·Accuracy²/Power* product of the basic voltage and current circuit blocks is fixed by technological constants. This result is obtained because in the simple blocks the same transistor is responsible for the accuracy, speed and power specification of the circuit. The bandwidth limiting capacitor, for instance, is inversely proportional to the offset so that the *Speed·Accuracy²* product becomes independent of the transistor sizing. In this section we will further generalize these findings to general analog systems; we study in detail how the different performance specifications are related in a general feedback voltage processing system, implemented with an OTA. Similar results can be obtained for other types of systems.

In figure 2.12(a) a schematic for a voltage processing system is shown. The output voltage is fed back in series to the input so that a high input impedance is obtained, which is desirable in voltage processing systems [Gra 84, Lak 94]. The power consumption of this system is determined by the current consumption of the OTA; the bias current of the OTA is proportional to the required g_{mIN} of the input stage, which is on its turn dependent on the wanted gain-bandwidth product of the system. General feedback theory and general opamp/OTA theory [Lak 94, Gra 84] show that in a system as in figure 2.12(a), the gain in closed loop A_{CL} times the bandwidth in closed loop BW_{CL} is a constant called the gain-bandwidth product GBW. An expression for the GBW of many types of OTA's is [Lak 94, Gra 84]:

$$\text{GBW} = A_{CL} \text{BW}_{CL} = \frac{g_{mIN}}{2\pi C_d} \qquad (2.79)$$

where C_d is the capacitor that is associated with the dominant pole of the open-loop transfer function of the OTA and g_{mIN} is the transconductance of the input stage. If we assume an input stage operating in strong inversion and we derive the following expression for the power consumption by using (2.58) and (2.79):

$$P = \pi A_{CL} \text{BW}_{CL} C_d V_{DD} (V_{GS} - V_T) \qquad (2.80)$$

The relative accuracy of the system is expressed by (2.59). The offset voltage of the OTA is in first order determined by the area of the input devices and consequently is related to the input capacitance as is shown in (2.101) derived in section 2.5.1. The output swing is limited to V_{DD} so that the maximal input signal is $V_{inRMS} = V_{DD}/(2\sqrt{2}A_{CL})$ and the relative accuracy is then given by:

$$\text{Acc}_{rel}^2 = \frac{V_{DD}^2 C_{in}}{24 A_{CL}^2 C_{ox} A_{VT0}^2} \qquad (2.81)$$

and we can rewrite the power consumption as:

$$P = 24\pi \frac{(V_{GS} - V_T)}{V_{DD}} C_{ox} A_{VT0}^2 A_{CL}^3 \text{BW}_{CL} \text{Acc}_{rel}^2 \left(\frac{C_d}{C_{in}}\right) \qquad (2.82)$$

We obtain a relationship between the specifications of a general voltage processing system implemented with an OTA very similar to the relationship obtained for the basic voltage processing stage in (2.61). If a relation between the the input capacitance C_{in} and the capacitance of the dominant pole C_d exists, the trade-off between the different circuit specifications is, also for the general system, determined by technological and physical constants only.

When the stability of the system in figure 2.12(a) is studied it becomes apparent that in any OTA design the fraction C_d/C_{in} will be larger than or equal to 1. From the small signal equivalent for an implementation with an OTA in figure 2.12(b), the loop transfer function is derived:

$$T(s) = \frac{g_m R_1}{(1 + sR_2 C_L)(1 + sC_{in}R_1)} \quad (2.83)$$

using the following approximations:

$$A_{CL} = \left(\frac{R_2}{R_1} + 1\right) > 1 \text{ or } R_2 > R_1 \quad (2.84)$$

$$\text{and} \quad r_o \gg R_2 \quad (2.85)$$

Due to the presence of two capacitors in the system, we obtain two poles[†]:

$$f_a = \frac{1}{2\pi R_1 C_{in}} \quad (2.86)$$

$$f_b = \frac{1}{2\pi R_2 C_L} \quad (2.87)$$

To evaluate the position of these poles we need an estimation of the relative magnitude of the C_{in} versus C_L. The load capacitance of the voltage processing stage is determined by the input capacitance of the next stage. As is explained in section 2.5.1, the input capacitance of the second stage C_L will be smaller as the C_{in}, since the offset requirements for the second stage will be less severe. In order to meet a certain accuracy specification in a multi-stage system, we have to divide the allowed total offset over the different stages as can be concluded from (2.97); (2.101) indicates that the ratio of C_L/C_{in} will be proportional to $\sigma^2(V_{os1})/\sigma^2(V_{os2})$. A good design compromise is to make $\sigma(V_{os1})/\sigma(V_{os2}) = 1/\sqrt{A_{CL}}$, since then the offset of the total system is dominated by the offset of the first stage only. When we approximate the exact expression for A_{CL} in (2.84) by R_2/R_1, we derive the following relative position of the two poles:

$$f_b = \frac{f_a}{\sqrt{A_{CL}}} \quad (2.88)$$

[†] the voltage input source resistance is assumed to be zero: $R_s = 0$; a non-zero R_s is in series with R_1 for the calculation of the pole f_a. In (2.86) we have thus assumed that $R_1 \gg R_s$ and the location of the pole can be chosen by the designer. In the case that $R_s \gg R_1$, the location of f_a is fixed by R_s and the input capacitance C_{in}, so that no design freedom exists for the position of f_a. That case is covered in section 2.5.1 and equation (2.104).

so that f_b is the dominant pole and that C_L plays the role of C_d in (2.82). To assure a stable feed-back system the second pole $f_{nd} = f_a$ must be at least larger or equal than the unity-gain frequency of the loop-transfer function T(s) to have a phase-margin of at least 45 degrees so that we obtain the following condition:

$$\frac{f_{nd}}{\text{GBW}} \geq 1$$

$$\frac{1}{2\pi R_1 C_{in}} \frac{2\pi C_L}{g_m} \geq 1 \qquad (2.89)$$

$$\frac{C_L}{C_{in}} \geq g_m R_1 = T_{@DC} \overset{!}{\geq} 1$$

To guarantee stability the ratio C_L/C_{in} must be made larger as $g_m R_1$ which is the loop gain at low frequencies as can be concluded from (2.83) and which is always at least larger than 1. So to attain stability, we have to increase the load capacitance C_L since we cannot make C_{in} smaller due to the accuracy specification.

Due to stability requirements, the ratio of capacitors in (2.82) will always be larger than 1. Also for other design choices of $\sigma(V_{os1})/\sigma(V_{os2})$, stability requirements impose that $C_d/C_{in} \geq 1$. We can conclude that the power consumption of a voltage feedback system is thus larger than:

$$P \geq 24\pi \frac{(V_{GS} - V_T)}{V_{DD}} C_{ox} A_{VT0}^2 A_{CL}^3 \text{BW}_{CL} \text{Acc}_{rel}^2 \qquad (2.90)$$

Equation (2.90) leads to the following conclusions:

- the minimal power consumption is limited by the effect of transistor mismatch and the quality of the technology is expressed by $C_{ox} A_{VT0}^2$; and also the maximal attainable *Speed·Accuracy²/Power* product is limited by the input transistor;

- to optimize the total combined performance of a voltage processing system the input stages have to biased with a low $(V_{GS} - V_T)$ or in weak inversion;

- in open-loop stages[†] the minimal power consumption is proportional to the square of the Gain in (2.51), (2.63) and (2.66); when feedback is used the minimal power is dependent on the *cubic* of the closed loop gain (Gain) in (2.78) and A_{CL} in (2.90) due to the effect of the limited gain-bandwidth of an opamp or OTA and the stability requirements.

[†]although the simple processing stage in section 2.4.2.1 has feedback, its performance trade-offs are those of an open-loop stage since no second pole is taken into account in the calculations.

2.4.4 Circuit design guidelines

We have studied several basic circuit stages for current or voltage signal processing. In this section we summarize the conclusions of the different designs and we highlight the limitations of the applicability of the different trade-off relationships.

2.4.4.1 Current processing stages. From the design of the current amplifier it can be concluded that in order to obtain optimal *total performance* i.e. high speed, high accuracy and low power consumption, a current processing stage must be biased with a *high* $(V_{GS} - V_T)$ as long as V_T mismatches are dominant. However the $(V_{GS} - V_T)$ of the stage should not be increased above the $(V_{GS} - V_T)_m$ of the technology. The overall best performance is obtained when the current processing stage is biased at $(V_{GS} - V_T)_m$. In practical situations this will be impossible due to other specifications like, for instance, signal swing and limited power supply voltage.

2.4.4.2 Voltage processing stages. A voltage processing stage must be biased with a *low gate-overdrive voltage* $(V_{GS} - V_T)$ in order to obtain an optimal *total performance*; the lowest power consumption is obtained in weak-inversion. However the maximal attainable intrinsic speed in a transistor decreases for small $(V_{GS} - V_T)$; so the minimal $(V_{GS} - V_T)$ can be imposed by the required speed performance. In the next paragraph we illustrate this for the one-transistor voltage amplifier of section 2.4.2.1.

Limitations on the choice of $(V_{GS} - V_T)$. Lowering the $(V_{GS} - V_T)$ of a transistor in strong inversion, lowers its maximal cut-off frequency f_T which is defined as [Lak 94]:

$$f_T = \frac{g_m}{2\pi C_{GS}} = \frac{3}{4\pi} \frac{\mu(V_{GS} - V_T)}{L^2} \qquad (2.91)$$

Since the bandwidth and thus the maximal operating frequency of a voltage processing stage is proportional to the f_T of the transistors, see (2.56), lowering the $(V_{GS} - V_T)$ limits the maximal operating frequency of the circuit. The minimal gate-overdrive voltage $(V_{GS} - V_T)$, which guarantees a strong inversion behavior of the transistor is typically 0.2 Volts (appendix A). We define the maximal frequency that can be processed for a $(V_{GS} - V_T)$ of 0.2 Volts, as the corner frequency f_{coII}:

$$f_{coII} = f_T \mid_{(V_{GS}-V_T)=0.2} = \frac{3}{20\pi} \frac{\mu}{L^2} \qquad (2.92)$$

As long as the operating frequency is lower than the corner frequency f_{coII} and a minimal gate length is used, equation (2.63) for the calculation of the power consumption is valid.

For frequencies higher than f_{coII}, however, the $(V_{GS} - V_T)$ must be made proportional to the required bandwidth. When we substitute this result into (2.61) the minimal power consumption for the one transistor amplifier satisfies:

$$P = \frac{8\pi^2}{3} C_{ox} A_{VT0}^2 \frac{L^2}{\mu V_{DD}} \text{Acc}_{\text{rel}}^2 \text{BW}^2 \text{Gain}^2 \qquad (2.93)$$

The power consumption is now proportional to the square of the operating frequency !

Equation (2.91) suggests that the absolute maximal frequency response that can be achieved in a MOS technology is limited by the minimal length L and the maximal $(V_{GS} - V_T)$ that can be used. But the maximal speed of electrons in silicon is limited to v_{sat} ($\approx 10^5 m/s$) so that at high $(V_{GS} - V_T)$ biases, the transistor goes in velocity saturation [Lak 94]. The voltage current relation is not quadratic in $(V_{GS} - V_T)$ anymore; the transconductance g_m of a transistor with fixed dimensions does not increase with the current anymore so that the maximal cut-off frequency f_{comax} becomes a constant, independent of biasing:

$$f_{comax} = \frac{1}{2\pi} \frac{v_{sat}}{L} \qquad (2.94)$$

This is the absolute maximal frequency that can be achieved in a technology.

In section 2.4.2.1 the lowest power consumption is obtained for a transistor biased in the weak-inversion or sub-threshold regime. The maximal bias current I_{Mwi} that still biases the devices in weak-inversion is limited to [Lak 94]:

$$I_{Mwi} = 0.2\mu n C_{ox} \frac{W}{L} U_T^2 \qquad (2.95)$$

The maximal cut-off frequency in weak inversion is then derived by substituting (2.95) into (2.65) and is given by:

$$f_{coI} = \frac{n\mu U_T}{10\pi(n-1)L^2} \qquad (2.96)$$

so that for frequency requirements below f_{coI}, the transistor can be biased in weak-inversion and the power consumption is limited by (2.66) for a given accuracy and gain.

Using the technology parameters from appendix A, we calculate the values of the different cut-off frequencies for the 0.7 μm CMOS technology; the maximal frequency f_{coI} in weak-inversion is 240 MHz; the f_{coII} lies at 4.6 GHz, and

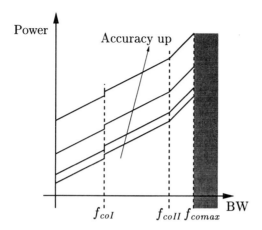

Figure 2.13. Evolution of the minimal power consumption as a function of the required bandwidth for the circuit, for different accuracies.

the absolute maximal frequency f_{comax} is 22 GHz. These numbers are very optimistic however, since in the first order analysis we have only taken into account the loading by the gate-source capacitances C_{GS}. In practical circuits, extra loads due to the e.g. the parasitic drain-bulk capacitors and interconnect or biasing sources parasitic capacitors reduce the frequency response of the circuits. Moreover, the derived expressions are for an open-loop amplifier, whereas in closed loop systems extra margins of safety have to be considered to assure system stability so that the maximal achievable bandwidth further reduces. Since the dominant parasitic load capacitors are typically also proportional to the transistor sizing (see discussion of α_{DB} in appendix A), the qualitative evolution of the power consumption discussed in this section remains valid for more complex circuits.

When the minimal power consumption for a one transistor voltage amplifier is plotted as a function of the required bandwidth BW, four operating regions can be distinguished, as is illustrated in figure 2.13:

- for BW $\leq f_{coI}$, the transistor can be biased in weak inversion and the power consumption is proportional to BW;

- for f_{coI} < BW $\leq f_{coII}$, the transistor can be biased in strong inversion with the minimal $(V_{GS} - V_T)$ of 0.2 Volts and the power consumption is proportional to BW but is the proportionality constant is about 5 times higher than in weak-inversion as derived in section 2.4.2.1;

- for $f_{coII} < \text{BW} \leq f_{comax}$, the transistor must be operated in strong inversion with a $(V_{GS} - V_T)$ proportional to the BW so that the power consumption scales with the square of BW;

- a $\text{BW} > f_{comax}$ is not achievable in the technology due to velocity saturation effects in silicon.

For all regions the power consumption scales with the square of the accuracy and the square of the gain. The discrete jump in the power consumption at f_{coI} is due to the limited accuracy of the transistor models we have used in the derivation of the minimal power consumption. We have considered a fixed limit between strong inversion and weak-inversion operation; however a smooth transition region called moderate inversion [Tsi 88] exists between the two regions so that also a smooth transition is observed in practical circuits for the power consumption; unfortunately simple and accurate hand-calculation models for the transistor behavior in this region are not available.

Limiting the value of $(V_{GS} - V_T)$ to obtain a minimal power consumption is thus limited by the required speed performance of the circuit block. The more demanding the speed requirements are and the more they approach the technology limits, the more excessive the power consumption becomes. This conclusion can be generalized for all voltage processing stages. In section 2.4.3, for instance, the bandwidth for a general OTA design in a feedback configuration is given by (2.79) and determined by the transconductance of the input stage and the dominant capacitor. As is shown, the dominant capacitor is larger than or equal to the input capacitance for stability reasons. The speed performance of the stage is thus limited by the intrinsic speed of the input device. The optimal power consumption is achieved for minimal $(V_{GS} - V_T)$ in the input stage, but also in that design reducing $(V_{GS} - V_T)$ can only be tolerated as long as the bandwidth specification allows it. The power consumption as a function of the required speed performance will have a similar characteristic as in figure 2.13.

2.5 IMPLICATIONS OF MISMATCH ON ANALOG SYSTEM PERFORMANCE

In the previous sections we have developed expressions for the trade-offs between the different circuit performance specifications of simple building blocks by studying their detailed design equations. From (2.51), (2.63) and (2.78) it is clear that a minimal power consumption is imposed for a given operation frequency (or bandwidth) and a given precision by the impact of device mismatch. This minimal power consumption is proportional to the matching quality of the

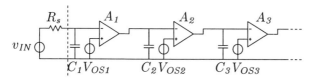

Figure 2.14. Schematic representation for a multi-stage voltage processing system.

technology which is expressed by $C_{ox}A_{VT0}^2$. In systems containing many building blocks the relations between the different specifications are more complex since more degrees of freedom exist. There is extra room for circuit and topology optimization but the transistor mismatch has basically the same impact. In this section we will derive relations for the power consumption as a function of the other specifications from basic general design equations. We show that a general relation can be derived for the minimal power consumption of a signal processing system due to the effect of the mismatch in the components.

2.5.1 General Multi-stage voltage designs

A general multi-stage voltage processing system is represented in figure 2.14. For a current signal processing system or a system architecture with currents and voltages intermixed similar schematics can be drawn and similar relationships can be derived. In 2.14 the offset voltage, input capacitance and gain of each stage is indicated. The relative accuracy - see (2.59) - of the total system is determined by the input signal RMS value and the equivalent input referred offset voltage which is calculated as follows:

$$\sigma(V_{OSeq}) = \sqrt{\sigma^2(V_{OS1}) + (\frac{\sigma(V_{OS2})}{A_1})^2 + (\frac{\sigma(V_{OS3})}{A_1 A_2})^2 + \ldots} \quad (2.97)$$

Mismatch effects put a lower boundary on the smallest signal that can be processed in a system. As a result, their influence is most important at the input stage where the signal levels are small. Therefore a designer will try to make the largest possible gain in the first stage A_1 so that the influence of the following stages on the accuracy specification is negligible[†]. The expression for $\sigma(V_{OSeq})$ is dominated by the term of the offset of the first stage so that the

---

[†]Since the power consumption of a block increases with the gain (see e.g. (2.63)), whereas the common required supply voltage for all blocks is determined by the wanted input swing and total gain, it is most power efficient to use as few stages as possible with the largest gain possible.

relative accuracy is approximately:

$$\text{Acc}_{rel} = \frac{V_{inRMS}}{\sigma(V_{OSeq})} \approx \frac{V_{inRMS}}{3\sigma(V_{OS1})} \tag{2.98}$$

The offset voltage of a stage is inversely proportional to the area of the input devices; the input capacitance is proportional to the area of the same input devices; thus a general relation between the input capacitance and the offset voltage exists. If we assume an input stage with a single transistor biased in strong inversion and since V_T mismatch is dominant, we obtain the following expressions:

$$C_{in} = 2/3 C_{ox} WL \tag{2.99}$$

$$\sigma^2(V_{OS}) = \frac{A_{VT0}}{2WL} \tag{2.100}$$

$$C_{in} = \frac{C_{ox} A_{VT0}^2}{3\sigma^2(V_{OS})} \tag{2.101}$$

For a stage with a differential input, the offset voltage is $\sqrt{2}$ times larger but the input capacitance is 2 times smaller so that the relation (2.101) between input capacitance and offset voltage remains valid. By applying (2.98) and (2.101), the input capacitance of the multi-stage system becomes:

$$C_1 \approx \frac{C_{ox} A_{VT0}^2}{3\sigma^2(V_{OSeq})} \tag{2.102}$$

The maximal speed of the system is of course determined by the bandwidth of the amplification stages; however, the pole generated by the input capacitor C_1 and the source resistance R_s is the upper boundary for the system speed:

$$f_{lim} = \frac{1}{2\pi R_s C_1}$$

$$f_{lim} = 3/2 \frac{V_{os1}^2}{2\pi R_s C_{ox} A_{VT0}^2} \tag{2.103}$$

From (2.98) it can be concluded that:

$$f_{lim} \text{Acc}_{rel}^2 \approx \frac{V_{inRMS}^2}{6\pi R_s C_{ox} A_{VT0}^2} \tag{2.104}$$

For the design of a multi-stage system the following conclusions are important:

- In high accuracy systems the best matching specifications have to be achieved in the *first or input stages* where the signal levels are the smallest; and the largest possible signal amplification has to be done as soon as possible.

- In a general design, (2.104) proves that the speed and the accuracy of the system are interdependent for a given input source impedance and their combination cannot be chosen freely but is dependent on the matching quality of the technology.

In this section we assumed that the input signal source drives the input capacitance of the signal processing system directly; then the power required for the signal processing is delivered by the input signal source. Many practical input voltage signal sources have relatively high output impedances, so that from (2.104) it is clear that an unbuffered input is not possible for high speed systems. The system schematic has to be changed so that the input capacitance is buffered for the input signal source and so that a high input impedance is obtained over a wide frequency range. Due to this extra buffering the power required to drive the input capacitance, is now delivered by the first block of the signal processing system itself, which we will study in the next section.

2.5.2 Limit imposed by mismatch on minimal power consumption

Mismatch is a random process and and its effect on the circuit behavior can be translated into time-invariant random DC error signals called offset voltages or currents. The random variations or the variance of the offset signals will be smaller if large devices are used since an averaging and smoothing occurs of the spatial errors sources responsible for the device mismatch. A certain accuracy will thus require a certain device area. However, this will lead to unavoidable parasitic capacitors at the input of the circuit proportional to the device area and thus inversely proportional to the offset signals as is demonstrated in (2.101).

When we want to buffer a certain voltage signal $v_S = V_s sin(2\pi ft)$ and drive it across a capacitor C, a current has to be delivered by the active part proportional to the voltage swing and the conductance of the capacitor at the signal frequency f: $i_C = V_s 2\pi f C cos(\omega t)$. In figure 2.15 a general schematic is represented for an active part driving a capacitor. Although no power is dissipated by the capacitor since its current and voltage have a 90 degrees phase-shift, for all active parts operating from a practical power voltage supply, power will be dissipated in the active part. The power consumption of the active part will depend on its efficiency which is a function of the mode of operation.

For a *class A operation* a DC bias is added to all signals equal to their amplitude. The minimum required supply voltage for the active part is V_{DD} −

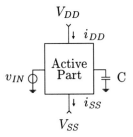

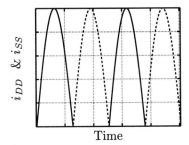

Figure 2.15. General schematic for an active circuit driving a capacitive load.

Figure 2.16. The supply currents for a class B active circuit driven with a sinusoidal input signal: (–) i_{DD} and (– –) i_{SS}.

$V_{SS} = 2V_s$, whereas the minimum required DC bias current is $I_B = 2V_s\omega C$ so that the following relation for the power consumption in class A is obtained:

$$P = 16\pi f C V_{sRMS}^2 \qquad (2.105)$$

When an active part operating in *class B* is used, no DC offsets are used but the active devices draw the signal current from the V_{DD} and deliver it into the capacitive load and they return the current from the load to the V_{SS}. The current signals i_{DD} from the V_{DD} supply and i_{SS} to the V_{SS} supply are depicted in figure 2.16. The average DC current flowing from the positive to the negative supply is $I_{DC} = (1/\pi)(2\pi f C V_s)$ so that the power consumption in class B is given by:

$$P = 8fCV_{sRMS}^2 \qquad (2.106)$$

Other classes of operating modes exist for the implementation of active systems, like class C, E, F, S; they have much higher efficiencies but can only be used for very specific types of signals and applications whereas class A and B are almost generally applicable. From (2.105) and (2.106) we can conclude that due to the presence of capacitors in the system a certain amount of power will be dissipated in the system to perform the signal processing at a certain speed or frequency f.

At this point we have all the relations available to estimate the power consumption of a system for a certain signal processing operation. From the definition of the dynamic range of a building block as the largest signal that it can process over the smallest signal, we have $DR = V_{sRMS}/(3\sigma(V_{OS}))$ and using (2.101) and (2.106) we obtain the following expression for the power consumption:

$$P = 24 C_{ox} A_{VT0}^2 f DR^2 \qquad (2.107)$$

This relation leads to the following conclusion: an analog signal processing system will at least consume the power expressed in (2.107) due to the effect of mismatch to perform a signal processing operation at a given frequency f and with a certain accuracy or DR. Equation (2.107) is the most general expression of the fact that the *Speed·Accuracy^2/Power* ratio is fixed by the technology mismatch quality.

2.5.3 Noise vs mismatch limits on minimal power consumption

Thermal noise and mismatch are both random processes and put a limit on the smallest signal that can be processed in a circuit; both physical phenomena will impose a minimal power consumption to achieve a certain DR specification and speed. In this section the minimal power consumption due to noise is derived and the limits imposed by mismatch and imposed by thermal noise are compared.

In [Vit 90b] the effects of thermal noise on the power consumption of a circuit is studied, which we repeat here for completeness. At a node with a capacitance C and driven by an impedance R, the total integrated thermal noise is:

$$V_{nRMS}^2 = \frac{kT}{C} \qquad (2.108)$$

with k the Boltzmann constant and T the absolute temperature. The dynamic range of the building block is given by $DR = V_{sRMS}/V_{nRMS}$, so that the minimal power consumption to achieve a certain DR imposed by thermal noise is given by:

$$P = 8kT f DR^2 \qquad (2.109)$$

To be able to compare these fundamental limits to the performance of realized circuits which all have different operating frequencies, not the power

66 ANALOG VLSI INTEGRATION OF MASSIVE PARALLEL SYSTEMS

Table 2.4. The minimal energy per cycle imposed by mismatch and by noise for a dynamic range of 1.

Technology λ_T	Type	Mismatch $24C_{ox}A_{VT0}^2$ [fJ]	Noise $8kT$ [fJ]
$2.5\mu m$ [Pel 89]	nMOS	4.3e-2	3.3e-5
	pMOS	-	3.3e-5
$1.2\mu m$ [Bas 95]	nMOS	2.12e-2	3.3e-5
	pMOS	-	3.3e-5
$0.7\mu m$	nMOS	6.3e-3	3.3e-5
	pMOS	-	3.3e-5

consumption but the energy per cycle P/f is evaluated as a function of the DR. In figure 2.17 the noise limit and mismatch limit for the minimal energy consumption are plotted as a function of the DR. The mismatch limit is technology dependent, since the product $C_{ox}A_{VT0}^2$ is a technology parameter and not a physical constant like kT. For present-day sub-micron CMOS technologies the limit on power consumption imposed by mismatch is about two orders of magnitude more important than the limit imposed by thermal noise, which can also be concluded from the values in table 2.4.

Analog Filters. In a high order analog filter the minimal power consumption per pole of the filter by the impact of noise is given by (2.109). In figure 2.17 the high dynamic range analog filter circuits referenced in [Vit 90b] are also represented. The dynamic range of filters is in first order not sensitive to matching or offset voltages. Mismatch will mainly influence the accuracy of the filter coefficients, the distortion performance and the power supply noise or common mode rejection characteristics of the filter which do not directly influence the dynamic range. So filter realizations only optimized to low power and high dynamic range or distortion, can consume less than predicted (2.107).

High speed A/D converters, are a better benchmark because offsets or mismatch limits the bit accuracy directly. So the minimal power consumption of A/D converters is clearly limited by mismatch. In figure 2.17 the energy per cycle of several A/D converters is represented. The high accuracy architectures have typically lower speeds and include digital or analog error correction circuitry, so they are only about 2 orders of magnitude from optimal perfor-

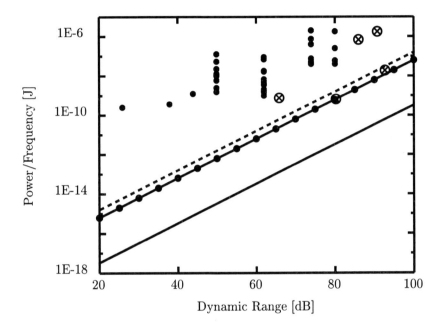

Figure 2.17. Comparison of the impact of thermal noise and mismatching on the power consumption of analog systems: (–) noise limit; mismatch limits: (– –) 1.2 μm and (–o–) 0.7 μm CMOS; Realized Filters ($\otimes$) [Vit 90b] and A/D converters ($\bullet$).

mance. The lower bit converters use very high speed architectures and cannot rely on error correction so their performance is 2 to 4 orders from the optimal. The limit performance takes only the consumption of the input stage into account but in practical circuits the subsequent stages also consume considerable amounts of power. The large discrepancies are further also due to the extra power taken up by parasitics.

The derived performance limit caused by mismatch is of course only valid for converter architectures for which the accuracy relies on component matching. Many converter architectures, like e.g. $\Delta - \Sigma$ or algorithmic converters, exchange conversion speed for accuracy and are made insensitive to component matching at the cost of lower conversion speeds. Their performance is typically limited by noise [Dij 94]. Also in [Pel 94], the impact of transistor mismatch on the power consumption of high-speed A/D converters is discussed. The analysis is based on a different calculation path but the predicted minimal power consumption is of the same order of magnitude as our results.

Summary. From the results in this section we conclude that in high speed, high accuracy analog systems, the effect of transistor mismatch imposes a higher minimal power consumption than the effect of thermal noise on the circuit specifications. It is important, however, to clearly understand the assumptions that lead to this conclusion.

Thermal noise is a fundamental physical limit to the minimal signal energy that can be used in electrical signal processing systems that arises from statistical thermodynamics [Mei 95]. Combining this limit with the best possible efficiency of a circuit, leads to the limit of (2.109). This limit cannot be broken by any circuit unless a more efficient architecture becomes available to transfer energy to a capacitive load from a DC power supply or other types of power supplies become available.

The origins of transistor mismatch on the other hand, are linked to the device structure and device physics and to the fabrication technology of integrated circuits; device mismatch originates from the stochastic nature of physical processes used for the fabrication like ion-implantations, diffusions or etching; the device structure using a channel in a doped material and its operation by modulating the channel resistance result in random fluctuations of the device properties and operation. For integrated circuit technologies as we use and fabricate them today, these physical limitations are very fundamental and device mismatch is unavoidable. In this perspective, the limits imposed by device mismatch, are restricted to signal processing systems realized using integrated circuits.[†] As such they are of course very important in the quest for

---

[†]In other integrated circuit technologies then CMOS like bipolar Si or GaAs, the same or similar fabrication principles are used. Fluctuations in the device operation or device mismatches also occurs and are governed by similar relations as those described in this chapter so that we can generalize this conclusion to all integrated circuit technologies.

minimal power consumption in integrated circuits but are of no importance for the fundamental physical limits of information processing.

As was demonstrated in this section, the mismatch limit is even more important than the noise limit in high speed analog systems. For high speed signal processing systems realized in present-day CMOS technologies the power consumption will be rather limited by the implications of device mismatch on the design and sizing of circuits to achieve a certain speed and accuracy specification than by the impact of thermal noise. However, when the system architecture or the application domain allows it, device mismatch can be circumvented by using offset compensating schematics as is discussed in section 2.6. As such the mismatch imposed limit is not important for all integrated circuits but unfortunately for a large part of high performance circuits.

2.5.4 Scaling of mismatch with technology feature size

2.5.4.1 Scaling of mismatch power limit. The minimal energy per cycle imposed by mismatch is proportional to $C_{ox}A_{VT0}^2$, as can be concluded from (2.107). The oxide capacitance per unit area is determined by the permittivity of silicondioxide (ϵ_{ox}) and the oxide thickness (t_{ox}):

$$C_{ox} = \frac{\epsilon_{ox}}{t_{ox}} \qquad (2.110)$$

For smaller feature size technologies proportionally thinner gate-oxides are used so that C_{ox} is technology dependent.

Variations in the fixed oxide charge, silicon-silicondioxide surface-states charge and the depletion charge are the main causes for threshold voltage mismatch [Pel 89, Miz 94, Laks86, Shy 84]. These variations in charge are transformed in threshold voltage variations by the gate capacitance. Therefore the threshold voltage proportionality constant A_{VT0} is expected to be proportional to the gate-oxide thickness t_{ox}. Experiments where the oxide thickness is varied and all other technology parameters are kept constant, are reported in [Miz 94]; the V_{T0} mismatch is indeed proportional to the gate oxide thickness and for a zero gate oxide thickness a zero mismatch is extrapolated.

In figure 2.18 the A_{VT0} of several processes with different feature sizes is plotted versus the gate-oxide thickness t_{ox} [†]; a linear relation is indeed observed between A_{VT0} and t_{ox} for technologies with a feature size down to 0.5 μm. A linear least-squares fit yields a slope of 0.5 $mV\mu m/nm$ and a non-zero intercept of 3.4 $mV\mu m$. The physical origin of the non-zero intercept is not yet clear. However the gate-oxide thickness is not the only technological parameter that changes with the feature size. For smaller line-widths and thinner gate-oxides

---

[†] I would like to thank Dr. M. Pelgrom of Philips Research in Eindhoven (NL.) for making available his measurement results reported in [Pel 89], which are also included in figure 2.18.

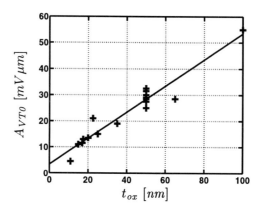

Figure 2.18. The experimental value of the threshold voltage mismatch proportionality constant A_{VT0} versus the gate-oxide thickness for different processes.

larger substrate doping levels (N_a) have to be used to avoid punch-through; larger doping levels increase the variance in the depletion charge and result in higher mismatching. A_{VT0} then becomes proportional to $t_{ox} N_a^{1/4}$ [Miz 94]. For a constant field scaling rule, the gate oxide thickness scales with $1/K$ and the substrate doping level scales with K so that the A_{VT0} scales with $1/K^{3/4}$. If the depletion charge mismatch is the dominant cause for threshold mismatch, and a constant field scaling can be assumed for the evolution of CMOS technologies, A_{VT0} is then expected to scale proportional to $t_{ox}^{3/4}$ when different processes are compared as in figure 2.18 and not linearly. This could be a possible explanation for the non-zero intercept when a linear fitting is erroneously assumed. However other explanations, justifying a non-zero intercept have also been proposed [Pel 89]. It is clear however that in the presented figure many parameters change from technology to technology and many different scaling laws are in use which could all have an impact on the A_{VT0}.

From the previous paragraph we can conclude that to evaluate the scaling of the technology mismatch $C_{ox} A_{VT0}^2$ we can assume an almost linear dependence of A_{VT0} on t_{ox}; the technology mismatch $C_{ox} A_{VT0}^2$ thus scales with t_{ox} and the matching behavior of transistors improves for smaller feature sizes. This is also confirmed by the experimental data in table 2.1. For the analog systems, a migration towards smaller line-widths should improve the circuit performance.

However, the maximal supply voltage also reduces for smaller line-widths [Mea 94, Hu 93], so that smaller signal levels have to be used. Especially in voltage processing systems, this results in a deterioration of the performance (see also [Pel 96]). If we examine equations (2.62) and (2.66), we conclude that

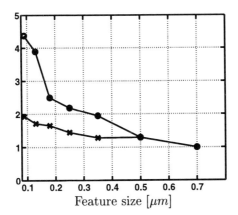

Figure 2.19. Scaling of the technology threshold matching quality as a function of the feature size; —o— does not take into account the decrease of the supply voltage for down-scaled technologies, whereas —x— accounts for the reduced quality of voltage processing stages due to the reduced available voltage swing in down-scaled technologies; the matching qualities are normalized to the value for the 0.7 μm technology.

the quality of the circuit will decrease if the supply voltage is reduced. Consequently the scaling advantage for the quality of voltage designs with smaller technology line-widths is reduced. In figure 2.19 the quality of sub-micron technologies, normalized to the quality of the 0.7 μm technology, is plotted versus the technology line-width using the process data from [Hu 93] and appendix A; the A_{VT0} is approximated as $0.5 \, mV \mu m/nm \cdot t_{ox}$. The —o— line uses $1/(C_{ox} \cdot A_{VT0}^2)$ as quality measure whereas the —x— line uses $V_{DD}/(C_{ox} \cdot A_{VT0}^2)$. The reduction of the supply voltage clearly reduces the scaling advantage of voltage processing stages.

For a current processing stage, the maximal signal swing is much less dependent on the available supply voltage. The voltage swing at the input depends only on the chosen bias modulation index. However, since the threshold voltage cannot be scaled proportionally to the supply voltage in order to limit the transistor cut-off leakage currents, the maximal $(V_{GS} - V_T)$ that can be used in a current processing stage reduces for lower supply voltages and the circuit performance will also degrade. However, in first order current processing stages are less sensitive to power supply voltage scaling.

Short channel effects also have a negative impact on the matching behavior as was demonstrated for the V_{T0} matching in section 2.2.3.E. Moreover, the increasing substrate doping levels in deeper sub-micron technologies make the parasitic drain to bulk and source to bulk capacitors relatively more and more important compared to the gate-oxide capacitance. This results in extra capacitive loading of the signal nodes and requires extra power to attain high speed operation.

We conclude that although the intrinsic matching quality of the technology improves for sub-micron and deep-sub-micron technologies, practical limitations make the theoretical boundary harder to achieve.

2.5.4.2 Relative importance of current factor and threshold voltage mismatches. At the start of the mismatch analysis we compared the relative importance of threshold voltage V_{T0} and current factor β mismatches on the behavior of transistors; for present-day processes the impact of the V_{T0} mismatch was clearly dominant. In the previous section the linear dependence of the A_{VT0} on the gate-oxide thickness was introduced so that the V_{T0} mismatch decreases for deeper sub-micron processes.

The proportionality constant A_β for the current factor has no clear relation to process parameters. The variation in β can be due to edge roughness, variations in the oxide thickness and mobility variations. The clear linear relation of the β mismatch with the gate-area excludes edge roughness as a dominant cause [Pel 89, Bas 95]; the low correlation with the V_{T0} mismatch excludes oxide variations as the dominant cause so that local mobility variations are the most probable dominant factor for β mismatches. In figure 2.20 the proportionality constants A_β are plotted for different processes[†] and we can conclude that the A_β is almost constant from technology to technology and has an approximate constant value of $2\%\mu m$.

When the scaling trends for A_{VT0} and A_β are compared, it is evident that the β mismatch gains in importance for deeper sub-micron technologies. This trend is confirmed by the values of the corner gate-overdrive voltage $(V_{GS} - V_T)_m$ in table 2.1 for different feature size technologies. For a slope in A_{VT0} with t_{ox} of $0.5\ mV\mu m/nm$ and a constant value of A_β of $2\%\mu m$ a value of $200\ mV$ is reached for $(V_{GS} - V_T)_m$ for a gate-oxide thickness of $4\ nm$; technologies with a feature size of about $0.25\ \mu m$ are expected to use this t_{ox} [Mea 94, Hu 93]. For this technology the β mismatch will be at least as important and even more important as the V_{T0} mismatch for the calculation of the accuracy of circuits in the whole strong inversion region.

The circuit design guidelines derived in section 2.4.4 are no more valid at that point. When β mismatch is dominant, current mirrors have to biased with the smallest possible $(V_{GS} - V_T)$ to obtain optimal total performance. But current mirrors can probably be biased at the optimal $(V_{GS} - V_T)$ of $(V_{GS} - V_T)_m$ - see section 2.4.1 - thanks to the low value of $(V_{GS} - V_T)_m$. At that bias point

---

[†]I would like to thank Dr. M. Pelgrom of Philips Research in Eindhoven (NL.) for making available his measurement results reported in [Pel 89], which are also included in figure 2.20.

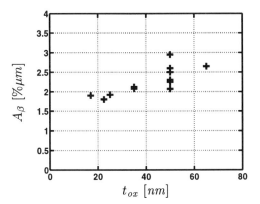

Figure 2.20. The experimental value of the current factor mismatch proportionality constant A_β versus the gate-oxide thickness for different processes.

the minimal power consumption of current processing stages for a given speed and accuracy is proportional to $C_{ox}A_{VT0}A_\beta$ from (2.52); this indicates that a further scaling of the technology would not further improve the performance but would result in a constant performance as far as the assumptions of a constant A_β and an A_{VT0} proportional to t_{ox} remain valid. The total performance of voltage processing stages as in (2.61) becomes proportional to the cubic of $(V_{GS} - V_T)$ when β mismatch is dominant over V_{T0} mismatch so that also the smallest possible $(V_{GS} - V_T)$ should be used. The minimal power consumption of voltage circuits then becomes proportional to $C_{ox}A_\beta^2$; this indicates that a further reduction of the oxide thickness would result in a worse performance for voltage processing circuits.

2.5.4.3 Summary. Thanks to the reduction of the V_{T0} mismatch with the oxide thickness, the move towards deeper sub-micron technologies will improve the intrinsic matching quality of the technology and should make lower power consumption possible. However, up to now no clear scaling of the β mismatch has occurred, so that for deep-sub-micron technologies of 0.25 μm and below, the current factor mismatch will become dominant; at that point further scaling of the technology is thought not to improve anymore the total performance of analog systems.

2.6 TECHNIQUES TO REDUCE IMPACT OF MISMATCH

Many techniques have been developed to reduce the impact of mismatches on the performance of analog building blocks. The time-invariant nature of the

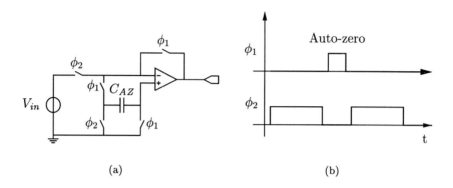

Figure 2.21. Auto-zero comparator schematic (a) to eliminate the effect of DC offset and (b) the timing for the switches.

offset signals caused by mismatches, allows the sampling of the error signals and subtracting them from the output. However, due to the limited availability of only accurate very short term analog memories, this compensation has to be repeated at regular time intervals during the system operation.

Auto-zero. In comparators using an auto-zero compensation technique, the offset voltage of the comparator is first sampled in a dead period and stored dynamically on a capacitor; this voltage is then subtracted from the input voltage before the comparison. A simple implementation of this scheme and the timing of the control signals is represented in figure 2.21 [All 87, Lak 94]. Similar compensation techniques have been used in A/D converters and switched capacitor circuits to improve the accuracy.

However several limitations exist for the applicability of this technique. First of all, dead periods in the system operation have to be available so that the analog building block can be disconnected from the signal processing chain and its offset can be sampled. Not all system architectures have dead periods available.

After the sampling of the offset, it has to be stored on an analog memory. For integrated circuits, storing an analog signal as a charge quantity on a capacitor is the most practical technique. Every analog memory only has a limited retention time due to unavoidable leakage currents. Especially in high accuracy applications, where a highly accurate storage is required, the retention time becomes short so that the memory has to be refreshed with short intervals; the continuous operation period of the system becomes small for high accuracy signal processing.

The switching of the capacitor in 2.21, which is necessary to reconfigure the system back into an operational mode, introduces additional errors due to clock-feed-through and charge injection [VPg 88, Vit 90c]. Via the overlap capacitors between the hold capacitor and clock lines, an error charge is put on the holding capacitor when the clock signals change. These errors can be compensated by using complementary switches and complementary clock signals but they increase the system complexity even more. When the switch-transistors are turned off, the mobile carriers that made up the conductive channel flow away to the source and drain so that a charge is injected onto the hold-capacitor; this introduces an error in the sampled offset voltage. Extra compensation switches can be used to reduce the charge-injection but again increase the complexity of the system [VPg 86]. The error reduction is limited by the matching between the switches and the compensation switches. Again transistor mismatch puts a boundary on the possible accuracy, but now it impact will be at least on order of magnitude less important.

Chopping. Chopper stabilization is another common technique for the reduction of the effect of offset voltages and 1/f noise on the accuracy of the signal processing [All 87]. The band-limited input signal of the amplifier is chopped by a multiplier driven with a 1/-1 signal, so that in the frequency domain it is modulated around the odd harmonics of the chopping frequency. This modulated signal is then amplified. The output signal is then again chopped so that is demodulated. The offset errors at the input of the amplifier are unaffected by the chopping and cause offsets in the output signal. The output chopping modulates these offset signals to high frequencies so that their effect in the passband is very small.

This techniques strongly reduces the effect of mismatch and 1/f noise. It requires however a high frequency operation of the amplifiers and the multipliers whereas the signal passband is only limited. Therefore, it is practically only useful for the design of high precision low frequency applications. In practical implementations a differential implementation of this effect is preferable in order to reduce the effects of power supply noise, charge injection and clock feed-through.

Trimming. Trimming techniques are typically applied after the fabrication of the integrated system. The accuracy of the system is evaluated and is corrected by changing certain component parameters. Resistor values can be changed with laser-trimming. Discrete trimming can be achieved by switching extra resistors or capacitors in parallel to trim the accuracy. This can be achieved with digitally controlled switches. When the calibration is done once, long term digital memory cells have to be available which require for instance the availability of EEPROM devices in the technology, which is far from standard. When a self-calibration scheme is applied, the component will calibrate itself at every power-up. However this requires on-chip references so that the system accuracy can be measured; the accuracy of the references must be larger than the required system accuracy so that their design can be very complex. On the other hand analog fuses can be used for a one-time calibration but they are also not standard components and increase the technology cost. The accuracy improvement will be dependent on the resolution of the switch-able elements which can lead to important extra area consumption. The most important disadvantage of off-line trimming techniques is the extra testing or calibrating time required after fabrication which increases the cost of the device substantially.

The implication of the offset compensation on the power consumption of the building blocks can be very important. Basically if ideal compensation is available, mismatch has not to be taken into account in the design and the mismatch limitation on the minimal power consumption is not important. In practice however, the accuracy specifications are relaxed and only the last bits or percents are achieved with compensation or trimming. This implies that these techniques reduce the limit imposed by mismatch typically by an order of magnitude. When the area and complexity increase of the system can be tolerated or the fabrication technology has extra trimming facilities, offset compensation techniques should be applied and will enable important power saving.

2.7 LINK WITH HARMONIC DISTORTION

Throughout this chapter we have focused on four main specifications of circuits and systems: accuracy, bandwidth or speed, gain and power consumption. However in many analog systems the linearity of the circuit or its distortion performance is of prime importance. In filter circuits, high linearity is necessary to avoid inter-modulation of unwanted signals into the passband or modulation of signal components. In all building blocks of telecommunication systems the distortion specifications play an important role in the design and performance trade-offs. To satisfy the linearity specification, the largest signal level in the circuit has to be reduced. Distortion has a direct impact on the maximal signals and thus influences the relative accuracy and dynamic range. Larger bias over signal ratio's have to be used so that more power is dissipated. In fact, when very high linearity specifications are to be met, the power consumption will be mainly originate from the impact of non-linearities [Wam 96b, Wam 96a].

In all calculations and derivations of this chapter we have made implicit assumptions about the allowed distortion. In the basic current amplifier (section

2.4.1) we assumed a bias current modulation index of 1/2 which has direct implications on the linearity of the amplifier. In the voltage processing circuits (section 2.4.2), the limitation on the maximal input signal was only limited by the maximal output swing; when linearity is important other constraints will exist and only a lower maximal input signal will be acceptable.

Since the distortion specification limits the maximal allowed signal, it has an impact on the relative accuracy of a circuit and consequently influences the quality of a circuit design. In voltage processing circuits, for instance, the relative accuracy is determined by the maximal input signal V_{inRMS} over the input referred offset signal V_{OS} (see also (2.59)):

$$\text{Acc}_{\text{rel}} = \frac{V_{inRMS}}{3\sigma(V_{OS})} \qquad (2.111)$$

The maximal input signal depends on the allowed output signal swing and the gain (Gain) of the block. The output distortion reduces the allowed swing at the output to a fraction of the maximal swing, which is typically half the power supply $V_{DD}/2$; similarly the input distortion can further reduce the allowed maximal input signal swing. The linearity specifications and the distortion thus reduce the V_{inRMS} to:

$$V_{inRMS} = \alpha_{disto} \cdot \frac{V_{DD}}{2\sqrt{2}\,\text{Gain}} \qquad (2.112)$$

The parameter α_{disto} depends on the linearity specification and the distortion performance of the circuit or building block and is smaller than or equal to 1. After substitution into (2.111) and (2.61), the quality of the one transistor voltage amplifier becomes:

$$\frac{\text{Gain}^2 \text{BW}\, \text{Acc}_{\text{rel}}^2}{P} = \alpha_{disto}^2 \cdot \frac{V_{DD}}{24\pi(V_{GS} - V_T)A_{VT0}^2 C_{ox}} \qquad (2.113)$$

The more relaxed the linearity specifications are the closer α_{disto} approaches unity and the better the total performance of the design becomes. For high linearity constraints the α_{disto} factor becomes small and as a result the power consumption for a given gain, speed and accuracy performance strongly increases due to the linearity requirements. For current processing stages a similar analysis can be done and the same conclusions can be drawn.

For the performance analysis of building blocks in this chapter, optimistic assumptions have been used in view of distortion performance and linearity specifications for the maximal attainable signal levels. In circuits that must meet high linearity specifications, the impact of distortion will strongly reduce the maximal signal levels and results in a lower dynamic range and a lower accuracy for the same current consumption and supply voltage. Consequently the linearity specification further increases the power consumption of the system to attain a specified speed and accuracy. In this perspective, the theoretical limits derived in this chapter, become even more difficult to approach in practical signal processing systems.

2.8 IMPLICATIONS FOR ANALOG PARALLEL SIGNAL PROCESSING SYSTEMS

Accuracy specifications. The results and analysis presented in this chapter are very important for the design of analog parallel signal processing systems. In their VLSI realization we want to achieve a high density combined with a low power consumption. Both objectives are influenced by the implication of transistor mismatch. The area of the circuits and thus their density is determined by the accuracy specification; but also the power consumption for a given speed is determined by the accuracy specifications since the performance ratio *Speed·Accuracy²/Power* is fixed by technology constants. A good accuracy specification of analog signal processing is thus very important; an over-specification results in poor speed and power performance and too loose accuracy specifications will result in faulty system operation. Therefore in the next chapter the generation of good accuracy specifications for the building blocks is treated in detail and the necessary theoretical evaluation methods are derived.

Since the area of a circuit is proportional to the Accuracy² · A_{VT0}^2, we can rewrite the ratio *Speed·Accuracy²/Power* as:

$$\frac{\text{Speed}}{\text{DensityPower}} \propto \frac{1}{C_{ox}} \qquad (2.114)$$

so that we clearly see that the total performance of an analog parallel signal processing system is also limited by a technology constant of the used VLSI technology.

Weak Inversion Operation. The study of the optimal design of basic building blocks shows that current processing blocks should be biased with large gate-overdrive voltages ($V_{GS} - V_T$) and voltage processing blocks should be biased with small gate-overdrive voltages and if possible in weak-inversion. This analysis thus shows that weak inversion operation of transistors is interesting for the VLSI implementation of parallel systems and neural systems but only for the voltage processing and the voltage mode blocks. Any current processing block must be biased in strong inversion to obtain good performance. An

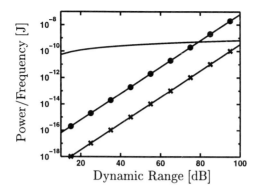

Figure 2.22. The energy consumption for a digital implementation (–) of a pole as a function of the required dynamic range; the minimal energy consumption of analog implementations due to the impact of mismatch (–o–) or noise (–x–).

OTA circuit for instance should be designed with its input transistors in weak inversion but the biasing and current mirror stages should use strong inversion as much as possible. A more correct statement is thus that for the implementation of analog parallel systems an optimal combination of weak and strong inversion biasing is to be used.

Analog or digital implementation. In order to decide whether analog or digital circuits are best suited for the implementation of massively parallel signal processing systems we have to compare their speed, power and accuracy performance or equivalently their power/speed ratio as a function of the dynamic range. In figure 2.22 the fundamental limits on the energy per cycle for analog circuits imposed by transistor mismatch and thermal noise are plotted; also the energy per cycle for a typical digital implementation of a single pole transfer function is plotted after [Vit 90b]. The power consumption curve of analog circuits has a steeper slope but starts from a much lower intercept; the curve for the digital circuits is only logarithmically dependent on the dynamic range, since increasing the dynamic range basically only requires the addition of an extra bit, but it starts from a much higher intercept.

Analog circuits are thus clearly advantageous for the implementation of systems with a low dynamic range requirement or low accuracy specifications. This type of specifications typically are required by massively parallel systems which perform perception tasks since they obtain their overall performance from the parallellism rather than from the high quality of the building blocks. Digital circuits are however clearly preferable when high dynamic ranges are required as in typical signal restitution tasks [Vit 90b, Vit 94]. However, transistor mismatch considerably lowers the dynamic range limit below which analog systems perform better compared to the limit imposed by thermal noise.

For the implementation of massively parallel signal processing systems analog circuits offer the best performance. However, the implications of transistor mismatch must be carefully taken into account; therefore good accuracy specifications are extremely important to obtain a high quality chip implementation.

2.9 CONCLUSIONS

This chapter discusses the implications of transistor mismatch on the design and on the performance of analog circuits and systems. First a characterization method is developed to extract the parameters for quantitative mismatch models. Since transistor mismatch is caused by statistical phenomena, a large number of sensitive measurements have to be performed using dedicated test circuits and a dedicated automatic measurement set-up. They show an inversely linear dependence of the parameter mismatch on the gate area of the transistors. When migrating to sub-micron and deep-sub-micron technologies, the influence of the short and narrow channel effects have to be accounted for in model extensions.

This firm quantitative understanding of transistor mismatch forms the basis for the analysis of its impact on circuit and system performance. We prove that the maximal total performance or the performance ratio *Speed·Accuracy²/Power* of elementary voltage and current building blocks is determined by the chosen biasing point – the gate-overdrive voltage $(V_{GS} - V_T)$– and the technology matching quality. A current processing stage must be biased with large $(V_{GS} - V_T)$ to obtain the best total performance whereas voltage stages must be biased with low $(V_{GS} - V_T)$'s as long as the bandwidth requirements allow it. Under optimal biasing conditions, the total performance or *Speed·Accuracy²/Power* of the stage is fixed and inversely proportional to the technology mismatching expressed by $C_{ox}A_{VT0}^2$; the lower the $C_{ox}A_{VT0}^2$ of a technology the better the circuit performance. This result thus explicitly states that a circuit designer can only trade one specification for another; but the best total attainable performance i.e. a combination of a high speed and a high accuracy at the same time as a low power consumption, is limited by the implication of transistor mismatch.

These results are then extended for more complex circuits, including opamps and feedback systems. Moreover, for a general analog signal processing system, we show that transistor mismatch again puts a fundamental limitation on the minimal power consumption for a given frequency and accuracy performance.

This technological limitation is even several orders of magnitude more important as the physical limitation imposed by the effect of thermal noise. For high speed and massively parallel analog systems and any other analog system where no offset compensation or calibration can be done, mismatch is the performance limiting effect, and the system design must carefully take mismatch into consideration.

The analysis of the scaling of the mismatch behavior with the technology feature size, shows that the evolution towards deep-sub-micron technologies improves the matching quality of the technology. However, the smaller power supply voltages in scaled-down technologies limit the available voltage swing and reduce the scaling advantage. The more pronounced short channel effects and velocity saturation effects also deteriorate the performance of analog circuits and reduce the scaling benefits. Furthermore, for very deep-sub-micron technologies ($< 0.25~\mu m$), the current factor mismatch prevent a further quality improvement so that further down-scaling will have no positive effect as far as the presently available data indicates.

At the end of the chapter, the implications of these results for the design of massively parallel analog signal processing systems are discussed. Due to the fixed ratio of performances, the importance of good accuracy specifications is demonstrated. The generation of accuracy specifications requires a sound understanding of the impact of random errors on the system level, a subject which is treated in the next chapter. Moreover, we show that analog circuits are indeed more power efficient for low accuracy levels than digital implementations, but mismatch lowers the boundary where digital implementations take over. We also indicate that the sub-threshold operation of transistors, which is generally believed to be the best regime for massively parallel analog systems, is not optimal for all circuit functions.

It is important to stress that the methods and analysis presented in this chapter are generally valid for the design of many types of analog systems. Transistor mismatch is an important factor in the design of analog circuits and therefore a good quantitative transistor mismatch model for the used technology should always be available to the analog designer. A good knowledge of mismatch allows a better optimization of the circuit design and avoids that very large safety margins have to be taken in the design which result in poor power and speed performance.

3 IMPLEMENTATION-ORIENTED THEORY FOR CELLULAR NEURAL NETWORKS

3.1 INTRODUCTION

To design analog VLSI circuits that implement the parallel processing systems developed by system theory researchers, the gap between the world of system theory and the world of analog circuit design has to be bridged. Both worlds use different conventions, use different resources, have different optimization goals and different constraints. In table 3.1 some of these differences are summarized. The theoretical descriptions of systems use a much higher level of abstraction than the circuit implementations. Translating the abstract description of a system as an algorithm or a block diagram using ideal primitive computing blocks, into an electronic function with the same behavior, requires a lot of extra information. The circuit designer is especially confronted with the limitations of the circuit blocks. A circuit implementation has extra unwanted poles in the transfer function compared to the system-level description; these poles can cause unexpected instabilities which make the circuit useless. The non-linear characteristics of electronic devices cause non-linear behavior of the blocks and limit the signal ranges that can be processed. The mismatch between the circuit components results in limited accuracy in the computations and spread of the

Table 3.1. Differences between the system theory and analog circuit implementations.

	SYSTEM THEORY	ANALOG CIRCUITS
RESOURCE	computers	silicon
TOOLS	mathematical equations	physical phenomena
QUANTITIES	variables	voltages currents charges
ABSTRACTION LEVEL	algorithms block diagrams	schematics circuit implementations
CONSTRAINTS AND LIMITATIONS	memory simulation time	area power/power-density package/no pins
PRIMITIVES	ideal elements - integrators - multipliers - PWL	circuit imperfections 2nd poles and zeros non-linearities, offsets, spread smooth characteristics

system characteristics; this can lead to malfunctioning of the system and as such influences the yield of the fabricated circuits.

In order to make a successful circuit realization of a new signal processing paradigm the circuit designer needs to know the sensitivity of the system to the typical imperfections of circuits so that he can derive specifications for the different building blocks that make up the system. In this area only limited information is available for the circuit designer from the system-level definitions or descriptions. To bridge the gap between the two worlds, extra implementation-oriented theory has to be developed that specifically deals with the implications of circuit imperfections on the system operation. Once the specifications have been derived, the circuit designer can evaluate different circuit topologies for their ability to implement the system. The circuit elements can be sized and if there are still degrees of freedom in the circuit design available, the circuit can be optimized towards low power consumption or high density for instance.

The main objectives for the design of analog computational circuits are:

- minimal circuit area or high cell density;
- high speed;

- and limited power consumption.

In chapter 2 it was shown that device mismatch causes a coupling between these specifications. Basically the *Speed·Accuracy² /Power* or the *Speed/Density·Power* of a circuit is limited by technological constants describing the matching quality of the technology and the necessary device area is proportional to the required accuracy in the computations. Since the *Speed·Accuracy² /Power* is fixed the accuracy specifications have very important implications on the speed and power consumption of the circuits. Therefore good accuracy specifications are indispensable for obtaining a chip implementation with a high performance.

For classical analog systems like filters, A/D or D/A converters and amplifiers specifications are determined by relying on linear system theory of continuous-time or sampled-data systems. The effect of the weak non-linearities in these circuits is accounted for by extensions to the linear system theory. In analog parallel processing systems as described in chapter 1, however, hard non-linearities are essential building blocks in the signal processing to enable the system to make decisions. The behavior and properties of analog parallel computational systems have to be modeled using non-linear system theory which has not many general applicable analysis methods.

In this chapter the impact of the circuit imperfections on the circuit operation of CNN's is studied. For all types of imperfections specifications are derived so that the circuit designer disposes of a full set of specifications for the cell building blocks. Section 3.2 reviews the different classes of imperfections that occur when a system description is translated in a transistor implementation.

In section 3.3 a method is developed to evaluate the robustness of cellular neural networks for random parameter variations in the connection weights or templates. It allows the prediction of the yield of a CNN for a given statistical variation on the template values. In section 3.4 this evaluation method is then used to derive accuracy specifications for the different building blocks in a CNN cell circuit. As such the method can also be used to do a template optimization towards higher robustness, so that more compact circuit implementations become possible.

Random errors do not only change the static behavior of the network but also influence its dynamics. Therefore the impact of random dynamical errors on the correct operation of the network is discussed in section 3.5.

In section 3.6 upper bounds on the maximal systematic static or distortion errors that can be tolerated without system performance degradation; they are important for the design of the output-non-linearity circuits, the template multiplier circuits and the cell resistor circuits. Parasitic time constants in the circuit implementation result in systematic dynamic errors in the CNN operation; they can cause faulty computations and in section 3.7 specifications for the parasitic time constants are derived.

86   ANALOG VLSI INTEGRATION OF MASSIVE PARALLEL SYSTEMS

Table 3.2. Classification of the main VLSI implementation imperfections by their error source type and their effect on CNN's.

	RANDOM	SYSTEMATIC
STATIC	offsets weight variations	non-linearities: distortion signal dependent behavior
DYNAMIC	time-constant mismatch	parasitic poles & zeros

In section 3.8 we evaluate the importance of noise signals for the circuit design and in section 3.9 we show that using a CNN for the realization of a linear filtering function or resistive grid function requires much harder linearity specifications.

We conclude the chapter with the introduction of a template redesign and template optimization method; the templates are optimized towards good hardware implementability and we show that with small modifications to the template values, very significant improvements in circuit performance is obtained.

3.2 INFLUENCES OF VLSI IMPERFECTIONS ON NETWORK OPERATION

The VLSI imperfections are subdivided in two groups depending on their effect on the computations. Two types of errors in the computation can exist: a *static error* occurs when the imperfections cause an incorrect cell behavior, independent of the evolution of its neighbors; or a *dynamical error* - comparable to races in digital circuits - which originate from internal dynamical errors in the cell or from an incorrect interaction between cells due to a large mismatch in the time constants.

A second criterion for subdivision is whether the error cause is systematic or random. Systematic errors are caused by deterministic processes and are the same for every cell. Their influence on the circuit behavior is derived using circuit theory. Random errors are caused by stochastic processes that occur, e.g., during the fabrication of the integrated circuits; no information on the amplitude of the errors is available only the standard deviation of the error can be computed; the errors are different from cell to cell or even from one building block in the cell to another identical building block. Their impact can only be evaluated by using the mathematical tools from statistics. The main imperfections of circuit implementations for the design of CNN's are tabulated in table 3.2 using these two criteria and are discussed in the next paragraph.

The non-linearity of the device characteristics of transistors results in a signal dependent behavior of the circuit blocks for large signal swings. Since this originates from the device characteristics it is a systematic effect and its main impact for CNN's is on the static characteristics of the cells. These distortion errors can be eliminated by correct biasing, good signal amplitude choices or non-linearity cancelation schemes like e.g. differential circuits topologies or trans-linear circuit techniques.

With every node of an integrated circuit a certain capacitance and a certain conductance is associated which causes a pole in the transfer function. Moreover, between any two nodes in the circuit there is always a certain coupling capacitance and a certain conductance which will result in a zero in the transfer function. Most of these poles and zeros will be located far from the operation frequency of the circuit or from its dominant poles and zeros so that they have very little impact. Unfortunately, in all practical circuits there are always more than one pole or zero that have a considerable effect on the circuit operation in the frequency range of interest. In a simple integrator block, for instance, a second pole can result in more complex dynamical behavior and can, for instance, cause instability in feedback loops. The cause of these imperfections is deterministic; as a result they have a systematic effect on the dynamic behavior of the CNN cells as can be concluded from their location in table 3.2.

The random variation in the device characteristics or the mismatching between identically designed devices cause random errors. The stochastic processes generating the component variations and device mismatches are time-invariant. This spread in component values results in a normal distribution of the effective template, capacitance and resistance values in the cells and limits the information representation accuracy or the computational accuracy. The spread in the template values can cause errors in the static behavior of the cells; the spread in cell capacitances results in mismatches between the time constants of the different cells which can introduce errors in the dynamical behavior of the cells. The absolute accuracy of the component values or the matching between devices can only be improved by using components with a large area which results in higher power consumption for the same speed performance as is discussed in chapter 2.

3.3 EVALUATION OF TEMPLATE ROBUSTNESS FOR RANDOM STATIC ERRORS

The robustness of templates and CNN's for the effect of random static errors is evaluated [Kin 94b, Kin 96a]. First the conditions for a correctly operating single cell is investigated. From these results the conditions for a correct operating multi-cell network can be derived and translated into accuracy specifications for the building blocks.

The CNN templates can be divided in two classes (see table 1.2). The *non-propagating templates* like the edge detector or noise filter, have a feedback template A with all zeros except for the self-feedback connection. Hence, the evolution of a cell is not dependent on the evolution of its neighbors. Each cell evolves independently and computes its output only from the input data. Consequently, a correct behavior of each individual cell *guarantees* a correct network behavior.

For the *information propagating templates*, like the connected component detector or holefiller, the cell's evolution depends on the evolution of its neighbors as the feedback connections in the A template have a non-zero value. The dynamics of the network are more complex. The activity of the network is, however, typically restricted to a small number of cells. In the connected component detector, for instance, at any point in time, one cell is in the middle of its state evolution, whereas its left neighbor is finishing its evolution and the right neighbor is just starting its evolution. This implies that if an individual cell behaves correctly and no races or dynamical errors occur, the network will behave correctly. The validity of this assumption will be proved with Monte-Carlo simulations in section 3.3.3. The correct operation of every single cell is still a *necessary* condition for the correct operation of the network.

3.3.1 Correct operation of a single cell

Dynamic routes. The behavior of one cell with its neighbors remaining in a fixed stable state, can be studied by drawing its dynamic routes for all possible combinations of its neighbors states [Chu 88a, Chu]. Since all neighbors remain in a fixed state, they all contribute a fixed signal in the state evolution equation. The cell state (x_c) evolution equation (1.1) can be rewritten as:

$$\frac{dx_i}{dt} = -G \cdot x_i + A_i \cdot f(x_i) + k \qquad (3.1)$$

where k contains the constant contributions of the I template, the B template and the A template with the self-feedback (A_i) excluded. In a dynamic route graph the state of the cell is represented on the x-axis and on the y-axis the corresponding time derivative of the state is plotted (see e.g. figure 3.1). For a state with a positive time derivative the state variable will grow to higher values which the arrows on the route indicate; for a state with a negative time derivative, the state variable will decrease. States with a zero time derivative are equilibrium points. Depending on the direction of the dynamic routes around the equilibrium point the stability of the equilibrium point can be determined.

IMPLEMENTATION ORIENTED THEORY FOR CNN'S 89

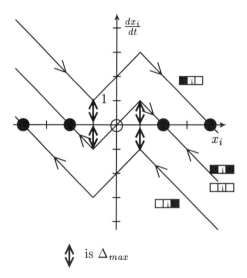

Figure 3.1. The dynamic routes for a cell i in the connected component detector for the different states of its two neighbors; the stable equilibria are indicated with a filled dot and the unstable with an open dot; Δ_{max} is defined on page 90.

In figure 3.1 the dynamic routes for a single cell in a connected component detector are displayed. Four different situations can be distinguished:

- when the left neighbor is in a high stable state - with an output of +1 or black - and the right neighbor is in a low stable state - with an output of −1 or white -, the cell always evolves to state of 4 with a high output, independent of its initial state; for a small perturbation from this equilibrium point, the cell is pulled back so that the equilibrium point is stable;

- when the left neighbor is white and the right neighbor is black, the cell evolves to the equilibrium with a state of −4 and has a white output; the arrows on the dynamic route point to the equilibrium point so that it is a stable equilibrium point;

- for the case of two black neighbor cells, the final state is dependent on the initial state; for a positive initial state, the cell evolves to a stable equilibrium point at 2 whereas for a negative initial state, the cell evolves to the stable equilibrium point at −2; as can be concluded from the direction of the dynamic routes the equilibrium point at 0 is unstable;

- when both neighbors are white the same dynamic behavior as for the case of two black neighbors is obtained, since the dynamic route coincides with the dynamic route for two black neighbors described in the previous paragraph.

Template value deviations. In figure 3.2 the influence on one of the dynamic routes of the connected component detector of a deviation in the different template values is shown. The deviation Δk in the k value results in a uniform shift of the dynamic routes. The Δk can originate from a deviation in the A (except A_i), B or I template values or in the output levels of the neighbors. A deviation of the cell conductance value G affects the slope of the dynamic routes, whereas a change ΔA_i in the A-template self-feedback coefficient results in a slope change in the linear region ($-1 < x_i < 1$) and a constant shift in the saturation regions ($x_i \geq 1$ or $x_i \leq -1$). In figure 3.2 the effect of positive values for the Δ's is shown but since the Δ's are random variables they can have negative or positive values.

The effect of these deviations on the correct cell behavior can be evaluated from the change in the equilibrium points of the cells for the different neighbor's states. From figure 3.1 and 3.2 it is clear that the dynamic routes can shift without a drastic change in the cell behavior as long as the shift (Δ_{route}) at the unit state 1 or -1 is smaller than the maximal allowed deviation (Δ_{max}), which is indicated in figure 3.1. For the connected component detector, for instance, the Δ_{max} is 1.

IMPLEMENTATION ORIENTED THEORY FOR CNN'S 91

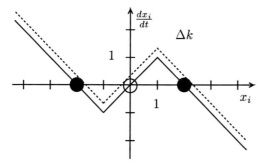

(a) The influence of a deviation in the k value, which can originate from an error in the A, B or I template values or the output levels of the neighbors

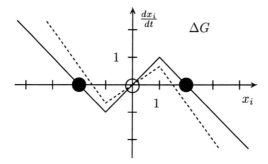

(b) The influence of a deviation in the cell conductance value

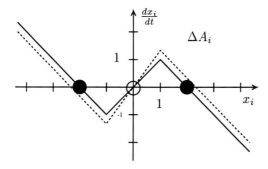

(c) The influence of a deviation in the self-feedback value in the A template

Figure 3.2. Influence of template coefficient variations on the operation of a CoCoD.

The variation of the dynamic route is dependent on the the variation of the template values which have a normal distribution since the circuit components are also normally distributed (chapter 2). Moreover, since the different template coefficients are implemented by different circuits and the variation of transistors is uncorrelated, the deviations will be statistically independent. The total shift of the dynamic route at the unit state is then calculated from (1.1) using (2.11):

$$\Delta_{route} = -\Delta G + \sum_c \Delta A_c + \sum_c \Delta B_c + \Delta I \qquad (3.2)$$

The standard deviation $\sigma(\Delta_{route})$ of the shift in the dynamic route at the unit state can be calculated from the standard deviation of the template values with (2.12), since the Δ's are normally distributed and independent random variables:

$$\sigma^2(\Delta_{route}) = \sigma^2(\Delta G) + \sum_c \sigma^2(\Delta A_c) + \sum_c \sigma^2(\Delta B_c) + \sigma^2(\Delta I) \qquad (3.3)$$

A formal definition for the maximal allowed deviation Δ_{max} is derived from (3.2) as follows:

$$\Delta_{max} = \min_{y_c, u_c \in \{-1, 1\}} |(A_i - G)y_i + \sum_c A_c y_c + \sum_c B_c u_c + I| \qquad (3.4)$$

The probability that a cell is operating correct $P_{correct}$ with the given deviation in the template coefficients is the probability that the deviation of the dynamic route in the unit state Δ_{route} is smaller than the maximal allowed deviation Δ_{max} for the given template. Since the circuit parameters have a normal distribution with mean 0 (see chapter 2), the Δ_{route} has a normal distribution with mean 0 and standard deviation $\sigma^2(\Delta_{route})$ so that the probability $P_{correct}$ is calculated from [Pap 91]:

$$P_{correct} = P(|\Delta_{route}| < \Delta_{max}) \qquad (3.5)$$

$$P_{correct} = 2 \cdot \text{erf}(\frac{\Delta_{max}}{\sigma(\Delta_{route})}) \qquad (3.6)$$

$$\text{erf}(x) = \frac{1}{\sqrt{2\pi}} \int_0^x \exp(-t^2/2) dt \qquad (3.7)$$

Up to now, the influence of variations in the template values on the behavior of a single cell has been determined using the technique of the dynamic route graphs. This allows the calculation of the probability that a single cell functions correctly. This information is now used to calculate the behavior of a large CNN.

3.3.2 Yield of a N-cell CNN

As discussed in section 3.3, the correct operation of every cell is a necessary and sufficient condition for the correct operation of a network of cells in the case of a non-propagating template. For an information propagating template the correct operation of all cells is a necessary condition; we will assume for now it is also a sufficient condition and in section 3.3.3 Monte-Carlo simulation results demonstrate the validity of this assumption. When *all* cells have to operate correctly for a correct network operation, the probability Y that a network with N cells is correct, is:

$$Y = P_{correct}^N \quad (3.8)$$

where $P_{correct}$ is the probability that a single cell is functioning correctly. The yield of the fabricated network chips is also Y. In figure 3.3 the yield for a template with a Δ_{max} of 1 is plotted as a function of the standard deviation of the deviation of the dynamic route in the unit state $\sigma(\Delta_{route})$ for different network sizes. The larger the network, the smaller spread can be allowed for a given yield.

An interesting approximation of (3.8) is obtained as follows:

$$\begin{aligned} Y &= P_{correct}^N \\ &= \exp(N \log(P_{correct})) \\ &= \exp(N \log(1 - P_{incorrect})) \\ &\approx \frac{1}{\exp(N(P_{incorrect}))} \end{aligned} \quad (3.9)$$

since $\log(1-x) \approx -x$ for small x. This approximation is valid as long as $P_{correct}$ is close to 1, which is almost always the case.

From the circuit designer viewpoint it is important to remark that below a certain threshold in $\sigma(\Delta_{route})$ the yield of the network becomes very high in figure 3.3. As long as the $\sigma(\Delta_{route})$ remains below this threshold a successful chip implementation is obtained. For a 128x128 network for instance $\sigma(\Delta_{route})$ must be smaller than 0.2 Δ_{max} to obtain a yield close to 100 %. From this threshold the accuracy specifications will be determined that guarantee a high yield for CNN chips and thus an economic VLSI implementation.

Evaluation method. With the results of this and the previous section we have established a method to calculate the yield of CNN for a given template and variation on the template values, which is summarized as:

1. for the template the Δ_{max} is calculated using (3.4) or is derived by inspection from the dynamic routes of a single cell for all possible states of its neighbors and its input;

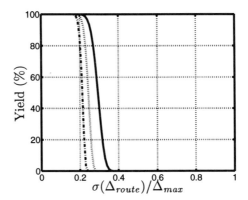

Figure 3.3. The yield of a network as a function of the $\sigma(\Delta_{route})$ — 32x32 network; ... 128x128 network and −.− 512x512 network.

2. from the standard deviation on the template elements the variance of the deviation in the dynamic route at the unit state $\sigma(\Delta_{route})$ is calculated using (3.3);

3. the probability that a single cell operates correctly $P_{correct}$ is calculated with (3.6);

4. the yield of the network or the probability of a correct network is calculated with (3.8).

Specification method. Inversely, for a given wanted yield of the network chips and a given size of the network N, the necessary $P_{correct}$ is calculated from (3.8); the allowed spread of the cell dynamic route $\sigma(\Delta_{route})$ is then obtained by inverting equation (3.6) for the given template and thus a given maximal allowed deviation of the dynamic route Δ_{max}. This spread is translated into the allowed standard deviation of the template values $\sigma(\Delta A_c)$, $\sigma(\Delta B_c)$ and $\sigma(\Delta I)$ using equation (3.3). This is used to derive the specifications for the different circuit components of the cell realization as is explained in section 3.4.

3.3.3 Monte Carlo simulation results

A deviation in the template coefficients has also an influence on the dynamic behavior or the state evolution of the cell, since the time derivative of the state changes, as is demonstrated in section 3.3.1. For the non-propagating templates this will only result in a change of the convergence time but will have no influence on the correct computations. However, when the information

Table 3.3. The necessary probability of correct operation of a cell ($P_{correct}$) calculated for a given yield and the corresponding simulated yield performance. The last column (σ_{error}) is the standard deviation of the simulated yield and the number of simulations is indicated between brackets. In the third column the allowed absolute deviation (σ_{abs}) on the template values is given.

	CALCULATIONS		MONTE CARLO SIMULATIONS	
Yield [%]	$P_{correct}$ [%]	σ_{abs}	Yield [%]	σ_{error} [%]
100x1 CONNECTED COMPONENT DETECTOR				
60.0	99.5	0.1785	57	6.8 (52)
70.0	99.64	0.1715	76.4	3.5 (140)
80.0	99.78	0.1635	81.4	3.1 (168)
90.0	99.89	0.1526	91.7	2.4 (150)
99.0	99.99	0.1286	98.9	0.78 (178)
32x32 MODIFIED HOLEFILLER				
95.0 (90.0)	99.99	0.0973	97.3	1.6 (150)
89.4 (80.0)	99.98	0.1022	87.4	2.7 (151)
4x1 CONNECTED COMPONENT DETECTOR				
70.0	91.50	0.29	66	2.9 (250)
80.0	94.60	0.26	78	2.5 (250)
90.0	22.50	0.225	87	2.5 (250)
99.0	99.75	0.165	99.6	0.6 (250)

is propagating through the network, dynamical effects can influence the correct computations. The evaluation method only takes into account the effect of static errors on the correct behavior of a CNN. To evaluate if the static effects of the template deviations are dominant over their dynamic effects, Monte-Carlo simulations of CNN's were done for information-propagating templates by randomly varying the template values.

In the left part of table 3.3 the wanted yield, the corresponding probability $P_{correct}$ and the corresponding relative accuracy of the template elements (cfr. section 3.4) are tabulated. Using this data, Monte-Carlo simulations of the CNN with random variations in the template values in the different cells are performed. As such the simulation results are an experimental verification of the accuracy of the yield estimation from the standard deviation of the template values for information propagation templates using equation (3.8).

DATA:	1. Network size
	2. Template & std. dev. on values
	3. Initial state of network and edges
	4. Input image of network and edges
	5. Correct output image
	6. Necessary evolution time
	7. Number of simulations: NS
ALGORITHM:	1. randomly generate templates for each cell
	2. transform network matrix in vector
	3. integrate states in time
	4. compare output image with correct output
	5. **if** correct **then** no_correct=no_correct+1
	else no_incorrect=no_incorrect+1
	6. **if** (number of simulations < NS) **goto** 1
	7. yield=(no_correct)/(no_correct+no_incorrect)

Figure 3.4. CNN Monte-Carlo simulation algorithm.

Simulation. The Monte-Carlo simulations are performed in MATLAB [Mat 92] using the algorithm presented in figure 3.4. The two-dimensional network is transformed into a one-dimensional vector by putting one row of cells, including the edge cells, after the other. A simulation starts with the generation of a different template for every cell with the internal MATLAB random generator for a normal distribution **randn** using the user-specified standard deviation of the different template elements. With the vector representation, the evolution of the network is calculated by integrating (1.1) from a specified initial state and input picture for a specified time-period by using a Runge-Kutta integration method implemented in the standard MATLAB function **ode23**. At the end, the end-state and output of the network is compared to the correct output and the correctness of the network is determined. This simulation is repeated until the specified number of simulations is executed and then the experimental yield of the network for the simulation run is calculated and reported to the user.

This simulation technique is far from the best possible implementation of a CNN simulator. Many more sofisticated simulators have been developed that are optimized for the simulation of CNN's [Dom 94]. By using a general tool like MATLAB, however, we have a much higher flexibility to implement different models or add additional effects at the expense, however, of longer simulation times.

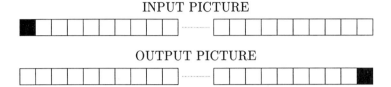

Figure 3.5. The input and correct output image used in the Monte-Carlo simulation of a 100x1 connected component detector.

CoCoD-template. The CoCoD-template is a template with a very complicated dynamical behavior. The inputs of the cells are not used and the input image is used as the initial state. During the evolution of the network, the original image is not stored in the network so that once errors occur, the network cannot recover and the final result will be incorrect.

For the CoCoD template the simulation is performed with the edges and all cells initialized to -1 (white) except the first cell; the correct output image is a single black cell at the right of the array (cfr. figure 3.5). For this test image all cells (except the first and the last one) have to make transitions from -1 to 1 and from 1 to -1, which implies that the correctness of all dynamic routes is checked. In figure 3.6 the state evolution of a correct 100x1 connected component detector is displayed. The variation in the template values over the cells results in a variation in the equilibrium points of the cells; this results in a spread of the final states of the different cells in figure 3.6. In table 3.3 the simulation result for a 100x1 CoCoD network and for a 4x1 CoCoD network for different yields are summarized.

Yield estimation accuracy. From the statistical point of view, the experimental estimation of the network yield Y using Monte-Carlo simulations is an experimental determination of the probability of an event [Pap 91, p. 251]. To estimate a probability we form a random variable x which can have two distinct values $\{0,1\}$ with a probability of respectively $\{1 - Y, Y\}$. The mean value of x is Y and its standard deviation is $\sigma^2(x) = (1 - Y)(Y)$. The determination of Y is thus equivalent to the determination of the mean of x.

We have performed n simulations of which n_s is the number of successful simulations. The point estimate for Y is $\bar{x} = n_s/n$. For large n, $\bar{x}$ is normally distributed with a mean Y and a standard deviation $\sigma^2(\bar{x}) = (1 - Y)Y/n$, so that an interval estimate of Y can be calculated for a given confidence level; the 95.5% confidence interval limits are $\{\bar{x} \pm 2\ \sigma(\bar{x})\}$ and the 99.7% confidence interval limits are $\{\bar{x} \pm 3\ \sigma(\bar{x})\}$.

98 ANALOG VLSI INTEGRATION OF MASSIVE PARALLEL SYSTEMS

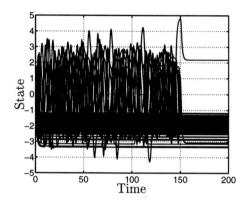

Figure 3.6. The simulation of a 100x1 CoCoD with varying template values over the cells for in input image [1 -1 ... -1].

The accuracy of the yield simulation is thus dependent on the number of simulations n executed. In the last column of table 3.3 the accuracy or $\sigma(Y)$ of each yield estimate is given. The 95.5% and 99.7% confidence interval estimates of the yield agree very well with the calculations.

Holefiller-template. For the modified holefiller template HOLE_MOD (see section 3.10 and table 1.2) the input image is set to a white image with black edges. All cells are initialized to black for this template. If the network operates correctly all cells remain black. If at least one cell becomes white due to errors, the whole image becomes white. With this test image, however, not all dynamic routes are tested. In fact, the network will only behave incorrectly if the shift in the dynamic routes is larger than Δ_{max} (=1) downwards. This is only the case for 1/2 of the incorrect networks. The $P^*_{correct}$ for this input picture is $1/2 + P_{correct}/2$ and the $P^*_{incorrect}$ is $P_{incorrect}/2$. Substituting these numbers in (3.9) leads to the conclusion that the yield obtained by the simulations is $\sqrt{Yield_{calculation}}$ so that for a specified yield of 90 % the expected simulated yield is 95 % and for a specified yield of 80 % the expected simulated yield is 89 % (see table 3.3).

Summary. The Monte-Carlo simulations in table 3.3 clearly show that the predictions of the network yield using the method outlined in section 3.3.2 on the basis of impact of the static random errors, agree very well with the experimental values obtained from the simulation of CNN's with randomly varying template parameters with the static and dynamic effects of these variations included in the simulations. We can conclude that the influence of the variations of the template values on the static behavior is dominant over their impact on the dynamic behavior for the tested templates.

3.4 GENERATION OF ACCURACY SPECIFICATIONS

In section 3.3 a procedure was introduced to determine the upper limit for the $\sigma(\Delta_{route})$ for a given network size, a given Δ_{max} of the template and a specified network yield Y. This limit can be translated into the accuracy specifications for the template circuits starting from equation (3.3), but the dependence of the random variation on the template value for the circuit implementation has to be known. Two cases are distinguished: the random variation in the template value is independent of the template value or the random variation is proportional to the template value.

Constant absolute error. For a template circuit implementation with a constant absolute error σ_{abs}, the random variation is independent of the template value. The maximal allowed σ_{abs} is derived from (3.3):

$$\sigma_{abs} = \frac{\sigma(\Delta_{route})}{M_{abs}} \qquad (3.10)$$

$$M_{abs} = \sqrt{\text{number of template values} \neq 0} \qquad (3.11)$$

if we assume that a zero entry in the template does not contribute any random variation in the dynamic routes. The validity of this assumption depends on the detailed circuit design of the template multipliers and is discussed in chapter 4.

Constant relative error. On the other hand, for circuits where the random variation in the template value is proportional to the template value, the circuit has a constant relative accuracy σ_{rel} independent of the template value. The σ_{rel} is calculated from the $\sigma(\Delta_{route})$ as follows:

$$\sigma_{rel} = \frac{\sigma(\Delta_{route})}{M_{rel}} \qquad (3.12)$$

$$M_{rel} = \sqrt{1 + I^2 + \sum_i A_i^2 + \sum_i B_i^2} \qquad (3.13)$$

Accuracy specifications. So depending on the circuit details, the circuit designer can generate the necessary accuracy specifications for the template circuits and the cell resistor by using (3.12) or (3.10). As an illustration, the necessary accuracies for several templates are listed in table 3.4 for a network of 128x128 cells with a yield of 90 %. The $\sigma(\Delta_{route})$ and M_{rel} and the

Table 3.4. The allowed $\sigma(\Delta_{route})$ calculated for a 128x128 network and for a yield of 90 %. From these specifications the maximal relative error σ_{rel} for the template circuits can be calculated from equation normalized estimate of the area of the synapses is given. In the last column the σ_{abs} is given.

Template	Δ_{max}	$\sigma(\Delta_{route})$	M_{rel}	σ_{rel} (%)	Area synapses	M_{abs}	σ_{abs}
AND	1.5	0.33	2.3	14	2× 51	1.7	0.19
CoCoD	1	0.22	2.6	8.4	3× 142	2	0.11
CoCoD_L	2	0.44	4.2	10.5	3× 91	2	0.22
EDGE	0.25	0.06	3.4	1.9	11× 2770	3.5	0.016
HOLE	0	0.00	5	0	∞	2.6	0
HOLE_MOD	1	0.22	5	4.4	5× 516	2.6	0.084
SHADOW	2	0.44	3.6	12.3	3× 66	2	0.22

corresponding σ_{rel} as well as the M_{abs} and σ_{abs} have been calculated. From these results some interesting conclusions can be drawn.

The HOLE_MOD template and CoCoD template have the same Δ_{max}, but the holefiller requires two times more accurate template circuits than the connected component detector, due to the higher number of connections between the cells. Hence, the more complex the template the more sensitive it becomes to mismatches. This can intuitively be explained: a cell must be able to distinguish a smaller signal change in a signal with a larger common-mode, which is also reflected in the higher M_{rel} value. This is known by analog designers to be a difficult task. The most complex template in the table, the edge detector EDGE, has indeed the strongest accuracy requirements.

By optimizing the template values, a template with a similar behavior can be obtained with a much higher robustness illustrated by the CoCoD_L – CoCoD_L is a connected component detector with larger template weights: A=[2,3,-2], B=0, I=0 – and HOLE_MOD templates. Moreover, some of the original templates like the holefiller HOLE [Mat 90] cannot be used in practical circuits due to their unrealistic accuracy requirements [Kin 94c]. The subject of template optimization is further discussed in section 3.10.

In chapter 2 the relative accuracy of a circuit was shown to be proportional to its area. In table 3.4 a normalized estimate for the area of the synapses in the cell circuit is also given. For a fixed function CNN chip implementation the total template circuit area is the number of non-zero template entries times the area of a template or synapse circuit which is inversely proportional to the allowed error σ_{rel}.

Programmable CNN implementations. For a programmable CNN implementation the worst-case specifications have to be taken into account so that the chip functions correctly for all possible templates, that can be programmed by the user. In figure 3.7, the necessary relative accuracy for template circuits with a constant absolute error is plotted for a set of templates for a 32x32 network and a yield of 90 %. A worst-case accuracy specification is extracted that guarantees a correct operation for all possible templates [Ros 94], and is indicated by the broken line in figure 3.7. For many individual templates this leads to a higher accuracy in the template circuits than necessary. This over-specification in accuracy leads to a higher power consumption, lower speed and lower cell density than in a fixed template chip implementation (cfr. chapter 2). Moreover, the implementation of programmable synapse or template circuit requires more area than fixed value synapses (cfr. chapter 4). Hence, programmable CNN chip implementations inherently have lower cell densities and higher power consumption.

Therefore, if a programmable system is desired for a specific application, the system or application designer has to decide with great care what level of programmability is necessary. As shown in the previous paragraph, the more flexible programmability is required, the lower the cell density and the higher the power consumption is.

3.5 RANDOM DYNAMICAL ERRORS AND THEIR IMPACT ON CNN BEHAVIOR

In this section we investigate the possible dynamic errors that can occur due to random variations in the cell behavior.

3.5.1 Non-propagating templates

A CNN programmed with a non-propagating template can be seen as a collection of cells that all evolve independently since their evolutions are not linked by any connection in the feedback template A. Therefore variations in the dynamical properties of the cells only results in a variation in the computational speed of the different cells and to obtain a correct solution, all cells especially the slowest one, have to be given enough time to converge to a stable equilibrium. When the evolution time of the network is fixed, there is always a certain probability that a cell has not converged before the end of the allowed evolution time so that an error in the network output can occur. This probability can be

102   ANALOG VLSI INTEGRATION OF MASSIVE PARALLEL SYSTEMS

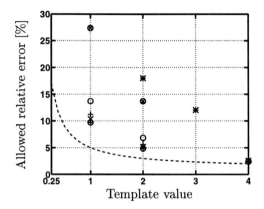

Figure 3.7. The allowed relative error for the template values for a 20x20 network and a yield of 95 %: o=CoCoD, x=HOLE_MOD, ⊗=SHADOW, ⊕=EDGE, +=NOISE, and *=PEEL templates (table 1.2). The specification for the programmable template circuit is (--).

easily calculated from the statistical properties of the time constants so that a safe estimation of the minimal template execution time can be made.

3.5.2 Propagating templates

First of all, the random variations in the template values including the cell conductance, result in a random variation of the cell time constant from cell to cell. Since a CNN can be a complicated system, including a lot of feedback loops within the cells and over two cells, one might expect dynamical errors to arise from these variations. But the Monte-Carlo simulations of propagating templates in section 3.3.3 have shown that the possible dynamical errors that occur are insignificant compared to the effect of the random variations on the static behavior of the cells. This conclusion is based on the templates investigated in section 3.3.3 and is, as such, in principle only valid for those templates. In the forthcoming paragraphs we will demonstrate that large classes of templates are indeed very insensitive to variations in the speed of computation of the cells determined by their time constants.

The propagating templates can be further subdivided into two groups: propagating templates with input or propagating templates without input.

Propagating templates with input. The holefiller and shadow generator are two examples of propagating templates with input. The A template has non-zero feedback connections and there are also non-zero entries in the B

template. The output of the network is dependent on its initial state and on the input picture presented to the cells. Since the input picture is stored into the network and remains available, a higher robustness can be expected, as the network can in principle still recover from its errors. We discuss the sensitivity of the two templates for dynamic errors in detail.

The *holefiller* operates in the following way. The input picture is presented to the cell inputs and all cells are initialized to black. The edges of the network are white. When the network evolution is started, the cells next to the edges start their evolution: when their input pixel is black or their four neighbors have a black output they remain black, otherwise they evolve to a white output. This behavior can be observed from the dynamic routes of a single cell for the different outputs of the neighbors and of its input in figure 3.9. Then the next layer of cells from the edges, start their evolution. The white information propagates in a wave from the edge of the network to the center. This process can be observed in figure 3.8, where subsequent snapshots of a holefiller network in operation are shown.

The question is now how sensitive this process is to mismatches in the time constants or computation speeds of the cells. If we assume that all individual cells operate correctly from a static viewpoint, – i.e. when they are tested individually with all possible states for their neighbors and input, they converge to a correct output – a cell in the network will only take the decision to evolve towards white, if indeed it is not part of a hole. Even if, for instance, one of its neighbors takes much longer to become white, once the neighbor is white, the cell will adapt its output, if necessary, and this information can propagate further through the network. Once a cell has evolved towards white from its black initial state, there is no way back, so that this situation has to be avoided at all costs. The correct static operation, however, guarantees that a cell will not make such errors. We can conclude that the holefiller template is completely insensitive to variations in the cell time constants, as long as the cells operate correctly statically.

The *shadow generator* template basically operates in a similar way and the same reasoning can be followed to demonstrate that the operation of this template is totally insensitive to variations in the cell time constants.

We can conclude that the propagating templates with input are insensitive to random variations of the cell time constants and that the yield of these networks is indeed only dependent on the static correct operation of all cells, discussed in section 3.3, which is also supported by the simulation results in section 3.3.3.

Propagating templates without input. A propagating template without input has the following properties: the A template contains connections to

104 ANALOG VLSI INTEGRATION OF MASSIVE PARALLEL SYSTEMS

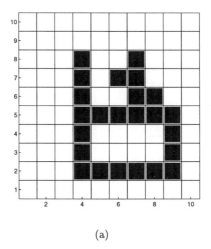

(a)

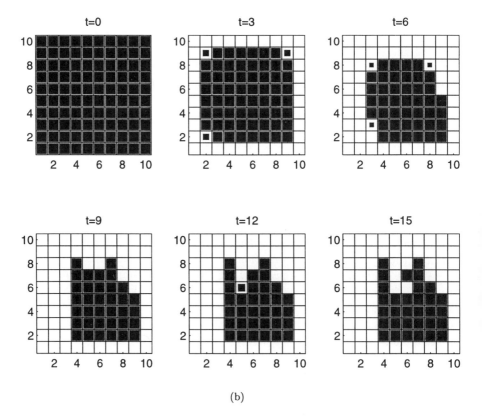

(b)

Figure 3.8. Subsequent snapshots in time of the evolution of the outputs (b) of a holefiller CNN showing the propagation of the 'white wave' from the edges towards the center of the network; in (a) the input image is represented.

IMPLEMENTATION ORIENTED THEORY FOR CNN'S 105

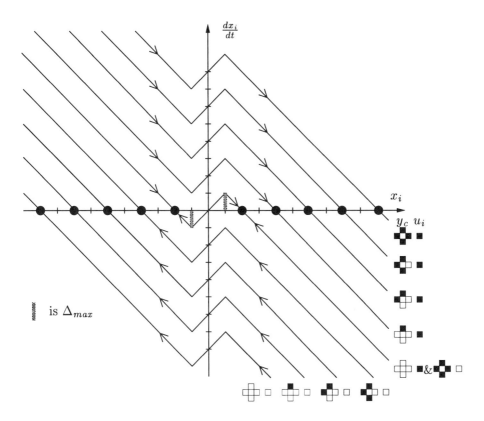

Figure 3.9. The dynamic routes of a cell in the holefiller CNN (HOLE_MOD) for the different fixed outputs of its neighbors and the possible input pixels.

the neighboring cells' outputs but the B template is zero so no feed-forward connections are used; the input image is used as the initial state of the network. During the evolution or computation, the original input is not stored, so that once errors occur in a temporary result, the information is not available anymore to recover from these errors. The class of the propagating templates without input inherently have a higher risk of higher susceptibility to dynamic errors.

A very interesting example of a propagating template without input is the connected component detector. This template looks simple, however, it results in very complex dynamics. The connected components in the image are reduced to a size of one pixel and at the same time these pixels move to a side of the image; this behavior is illustrated in the subsequent snapshots of the evolution of a 'horizontal connected component detector to the right' in figure 3.10. A pixel representing a component from the left side of the image, has to shift through several cells before it reaches its end-position. If during one of these shifts the pixel is lost, the network will not be able to recover from this error.

The random dynamical errors originate from a variation in the template elements, the cell conductance and the cell capacitance. In the Monte-Carlo simulations of section 3.3.3 the effect of the random variation of the template elements and the cell conductance have been taken into account. The simulation results in table 3.3 indicate that the static effect of these random variations is dominant over their dynamic impact. The random variation of the cell capacitance, however, has not been accounted for in the Monte-Carlo simulations.

The variance of the cell time constant is derived using (2.12), as follows:

$$\tau_s = R_s C_s \qquad (3.14)$$

$$\frac{\sigma^2(\tau_s)}{\tau_s^2} = \frac{\sigma^2(R_s)}{R_s^2} + \frac{\sigma^2(C_s)}{C_s^2} \qquad (3.15)$$

Depending on the chosen implementation, the cell resistor and capacitors are or fabricated in different process steps or implemented using different device properties so that their variations are always statistically independent and normally distributed. The worst case relative process variation of the absolute capacitance value of a capacitor is typically 20 % in a CMOS process so that its standard deviation is about 6.6 %. The allowed standard deviation on the conductance is dependent on the network size, the specified yield and the template; for the connected component detectors in table 3.3 the relative variation is significantly larger than 6.6 % even for high yields and large networks. Therefore the extra variation in the dynamic properties of the cells due to the variation of the cell capacitance is relatively small.

To evaluate the influence of capacitance variations on the correct behavior of the connected component detector, we have simulated 750 10x1 connected

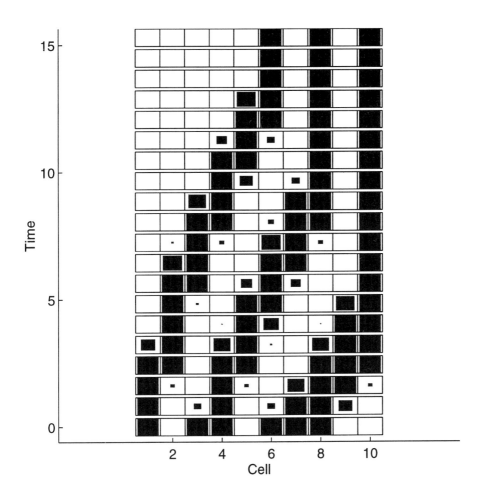

Figure 3.10. The evolution of the output of an 10x1 connected component detector with template A=[1,2,-1], B=0, I=0.

component detector networks designed for a specified yield of 99.9 % with the method in section 3.3; besides the random variation in the template elements, a random capacitance variation with a standard deviation of 6.6 % has been included. The simulation gives a 99.7 % confidence interval estimation of the yield of $\{99.7 \pm 0.12\ \%\}$. We can conclude that the extra variation in the time constant due to the spread in the capacitance value has indeed no significant effect on the yield of the networks.

Summary. The effect of random dynamical errors or mismatches in the cell time constants on the correct operation of CNN's is very small. For non-propagating templates and propagating templates with inputs, there is no direct dependence of the correct operation of the network on the time characteristics of the dynamics of the network. For propagating templates without input the effect of random errors on the correct static behavior is more significant than their effect on the dynamic behavior. We can thus conclude that random dynamical errors seem to present no significant problems for the investigated templates.

3.6 SYSTEMATIC STATIC ERRORS AND THEIR IMPACT ON CNN BEHAVIOR

To determine the influence of systematic static errors on the operation of a CNN, the same methodology as in section 3.3 can be used. The random and systematic static errors only differ in the nature of their causes but their impact is basically the same. We can thus evaluate the effect of a static error by controlling the correct operation of a single cell from its dynamic routes.

Non PWL non-linearity. In a practical circuit realization the non-linearity f, used to compute the output from the state, will never have a sharp piecewise-linear characteristic. Moreover, the realization of a good approximation to the PWL characteristic would require a rather complicated circuit; the implementation of a saturating non-linear characteristic is very simple as is discussed in chapter 4. For this type of non-linear function the shape to the dynamic routes changes slightly, as is illustrated in figure 3.11, but the basic behavior of the cell remains the same as long as the saturation level of the ideal and realized non-linear function are the same. In [Chu 88b] and [Hal 90] this insensitivity is also discussed.

Non-linearity in A template multipliers. Non-linearities in the circuit implementation of the A-template multipliers result e.g. in a reduction of the effective weight value with the output voltage. This non-linearity only

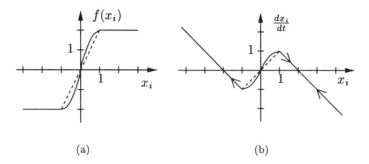

Figure 3.11. (a) The practical circuit realization of the output non-linearity f has a smooth characteristic (-) instead of the ideal piece-wise linear characteristic (--). (b) This has no significant influence on the dynamic routes of a CNN cell.

changes the slope of the scaled output non-linearity $A \cdot f(x_i)$ in the linear region slightly and has no influence on the behavior of the network as discussed in the previous paragraph. However, care has to be taken that the output signal in both saturation regions is correct. Therefore the reduction in the weight value can be compensated by using a tuning loop to derive the control signals of the template multipliers. The tuning loop adjusts the weight control signal so that the output signal in the saturation region reaches the correct value for the desired template. Since the effect is systematic and is present with the same amount in all cells, only a single tuning circuit for the whole network is necessary. This tuning strategy is further discussed in chapter 4.

Non-linearity in B template multipliers. When a non-linear template multiplier circuit is used for the B template, the feed-forward signals are not linear scalings of the input of the cell. This can introduce distortions in applications using gray-scale inputs for instance. For applications with binary inputs, the effect of the non-linearity of the template circuits can be compensated with a similar tuning circuit as for the A template circuits.

Cell resistor. In many advanced CMOS IC processes, no devices with a linear resistance are available, since the design of digital circuits does not require linear resistors. Several combinations of transistors are available to obtain a circuit with a resistive behavior, but the linearity of these devices is not very good (cfr. chapter 4). These non-linearities result in a change of the shape of the dynamic routes in all regions. In figure 3.12 the dynamic routes for a cell with a non-ideal cell resistor are represented. The qualitative behavior of the cells remains intact. The non-linearity of the cell resistor can also be taken

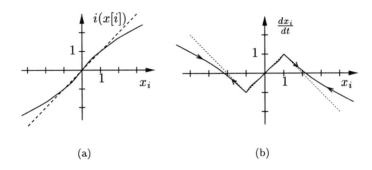

Figure 3.12. (a) A compact cell resistor circuit implementation has a non-constant resistance (-) and deviates from the ideal behavior (--). (b) The resistor distortion does not introduce a significant difference in the cell dynamic routes (-) compared to the ideal behavior (--).

into account in the weight tuning circuits as is discussed in chapter 4, by using the current through the non-ideal cell resistor at the unit state as the reference signal for the tuning of the A and B template signals. Even if linear resistors are available in the technology, the cell resistor is best used as a reference in tuning loop.

In very exceptional cases, the non-linearity of the cell resistance can be a problem if at the same time a large slope in the saturation region of the non-linear function circuit is present [Hal 90]. Due to the slope in the saturation region, the A template self-feedback signal is in part proportional to the state in all regions; this can result in a positive feedback in all regions so that the cells become unstable. This effect is dependent on the template settings and on the value of the cell resistance. However, to cause instability very large slopes are necessary. For the holefiller template, for instance, a slope of 33% is necessary for a unit cell resistor to make the cells unstable. In practical circuits the output resistance of the template circuits and non-linear function circuits are high compared to the cell resistance due to the relatively high output resistance of transistors so that this effect can be neglected.

Summary. We can conclude that the systematic static errors introduce no severe problems for the correct operation of a CNN or that effective measures to cancel their impact can be taken. CNN's do not require high linearity specifications for the cell building blocks; this leaves room in the design of a cell circuit to reduce the circuit complexity and as a result, to reduce the area and power consumption. To evaluate the impact of systematic static errors, the circuit designer can again simply evaluate their effect on the shape of the dynamic routes of a single cell.

3.7 SYSTEMATIC DYNAMIC ERRORS AND THEIR IMPACT ON CNN BEHAVIOR

Every node in a circuit has a (parasitic) capacitance and is driven by a source with a limited conductance, so that the node voltage cannot change instantaneously but will always require a certain time constant. In the CNN cell model, a 'large' time constant is introduced at the state node by the effect of the state capacitor and the state resistor. In a circuit realization, the capacitive loading of the output node is relatively large, since several template multipliers are tied to it. Consequently, a second pole and thus integrator is also present at the output node. In the rest of this section we refer to this pole at the output as the second pole . This pole introduces extra delays in the feedback loops created by the feedback template A; a loop exists in the cell due to the self-feedback and feedback loops, that span over several cells, also exist for propagating templates.

For a circuit designer, it is important to know how large the parasitic time constants can be relatively to the cell time constant, so that the circuit still functions correctly and remains stable. The circuit's parasitic time constants can be reduced but this typically costs extra power.

Non-propagating templates. In a CNN programmed with a non-propagating template, there is only a feedback loop within every cell, due to the necessary self-feedback to obtain a binary output. To investigate the effect of the second pole on the stability of the cells, we cannot use the classical linear feedback theory, due to the non-linear output function f. The behavior of the cell with two poles is modeled with the following state space equations:

$$\tau_s \frac{dx_i}{dt} = -x_i + A_i y_i + k \qquad (3.16)$$
$$\tau_o \frac{dy_i}{dt} = -y_i + f(x_i)$$

where x_i and y_i are the state and output of the cell, τ_s and τ_o are the time constants of respectively the state and the output, A_i is the self-feedback weight in the A template and k is the sum of the feed-forward signals. The equilibria of the cell are the pairs $\{x_i, y_i\}$ for which the left hand sides of (3.16) are equal to zero and for which the time derivatives are thus zero.

We must now investigate if the cell behavior is different due to the extra integrator that is present at the output of the cell. In most non-propagating

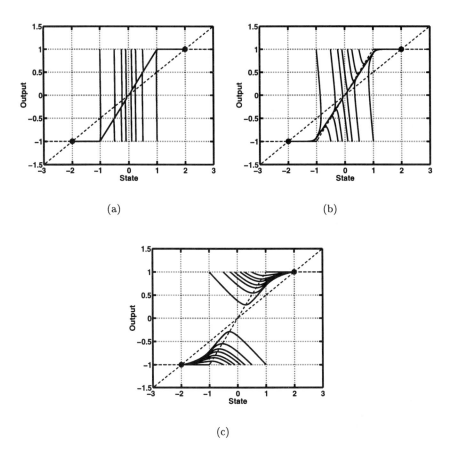

Figure 3.13. The state space trajectories (-) for the state x_i and the output y_i for different initial situations and different ratio's of the time constants: (a) $\tau_s/\tau_o = 100$, (b) $\tau_s/\tau_o = 10$, and (c) $\tau_s/\tau_o = 1$. The stable equilibria are indicated by o; they are at the crossing of the (- -) curves which represent the model equations in (3.16) for zero time derivatives.

templates, A_i is equal to 2. The most critical situation, then occurs when k is 0. Then the ideal dynamic route of the cell state x_i, for a zero time constant τ_o at the output, is the same as the ideal dynamic route represented in figure 3.2. So when no second pole is present, the cell state has two stable equilibria at -2 and at 2; for an initial state smaller than 0 the cell state evolves to -2 and for an initial state larger than 0 the cell state evolves to 2; at 0 an unstable equilibrium exists.

In order to study the effect of the second pole, we draw the state-space trajectories for the two-pole system in (3.16). We suppose the cell can only start with an initial value for the output of -1 or 1, which is the output of the previous computation. The state-space trajectories for different initial states and different ratios τ_s/τ_o are drawn in figure 3.13:

- for a τ_o which is 100 times smaller than τ_s, the behavior of the cell is almost ideal (figure 3.13(a)); the cell evolves to a final state of 2 in the initial state is larger than 0 and to a final state of -2 if the initial state is smaller than 0, *independent* of the initial value of the output node.

- For a τ_o 10 times smaller than τ_s, we see that for very small initial states, larger than 0, and for an initial value of the output of -1, the cell evolves to the equilibrium state at -2, which is different from the ideal behavior. For larger initial states the behavior is still correct (figure 3.13(b)).

- For a τ_o equal to τ_s, the cell state evolves to the equilibrium at -2 for all positive initial states if the output starting value is -1. At this point the behavior of the cell is incorrect for all positive initial states (figure 3.13(c)) ! Similarly an incorrect behavior occurs for the negative initial states.

The second pole indeed has an influence on the behavior of a cell in a CNN with a non-propagating template. Not only the initial state and the inputs of the cell and its neighbors, which is modeled by k in (3.16), but also the previous value of the output of the cell determine its final state. For most non-propagating templates, however, a fixed initial state of 1 or -1 is used so that they are less sensitive to this effect. We can conclude that, as a rule of thumb, the parasitic τ_o should be at least 10 times smaller than the wanted cell time constantτ_s and that the cell behavior will be close to the ideal behavior.

Propagating Templates. In a CNN programmed with a propagating template, feedback loops exist that span over two cells. The stability of the ideal CNN, including the feedback loops over two cells with two time constants, has already been studied for different templates and many results have been published [Chu 88b, Chu 90, Chu 92, Guz 93]. For a large part of the template space, stability or instability has been proven.

If we include the parasitic integrator at the output of the cell, we obtain a feedback loop with 4 poles and 2 non-linear functions f. Even more complex dynamic behavior as in the ideal case is now possible. The problem of the stability of this system is somewhat related to the stability of delay-type template CNN's [Ros 92], but there a fixed delay is used. A detailed study of the stability of these networks is very involved and is still a topic of intensive research by mathematicians and system theoretics.

If the parasitic time constant at the output is sufficiently small, we can assume the cell will almost behave as in the ideal case. Therefore, as a rule of thumb we will require to make the parasitic time constant 10 times smaller as the cell time constant. This approach is similar to the stabilization technique that is used in many classical feedback circuits with high order poles. Instability occurs if the poles approach each other, and a straightforward way of stabilizing the loop is by adding extra capacitance to make one time-constant considerably larger than the others. Then the loop behaves almost as a one pole system and is stable.

The specification of the relative position of the two poles, has important implications for the minimal power consumption of the circuits. It is clear that if better data is available to specify the relative position than the first order models used in this section, better specifications can be generated and better performing circuits are obtained.

3.8 NOISE IN CNN'S

In this chapter, thermal or $1/f$ noise have not been discussed as possible causes for random errors in CNN's. In chapter 2 the impact of noise on the accurate behavior of high speed analog systems is indeed shown to be orders of magnitude smaller as the impact of mismatch under the assumption that no offset cancelation techniques can be used. In CNN's we want to achieve a high cell density, so that the cell circuits cannot rely on offset cancelation techniques, due to their relatively high complexity. Therefore the impact of noise on the correct cell behavior can indeed be neglected. It is very unlikely that errors will occur in the operation of a CNN due to the effects of noise signals.

On the contrary, thermal noise has the beneficial effect that the network cannot settle into an unstable equilibrium state. Due to the presence of noise, small excursions around the equilibrium will always occur and will drive the network out of unstable equilibrium states. In some extended CNN models, extra noise sources are even introduced in the state equations [Chu 93].

3.9 CNN'S AS RESISTIVE GRIDS

The different analyses of this chapter have been made for the classical CNN's with a self-feedback in the A-template larger than the cell conductance, so that only stable binary outputs exist [Chu 88b].

CNN's can also be used in their linear region, if the self-feedback in the cells is smaller than the cell conductance. In fact, in [Shi 92] the implementation of a linear resistive grid by a CNN in the linear region is demonstrated. The state and the outputs of the cells remain in the linear region of the ideal PWL nonlinearity f. The effect of the different circuit imperfections, that are discussed in this chapter, on the behavior of linear grids has been covered partly in [Shi 92].

As such, this is an interesting extension of the processing capabilities of CNN's. Resistive grids are indeed used in many image processing systems [Mea 89, Kob 90, Yu 92]. The implementation of a resistive grid by the CNN, however, heavily depends on the linear dependence of the output on the state for states between -1 and 1. Consequently, from the circuit design perspective, using a CNN as a resistive grid introduces extra constraints for the linearity of the output non-linearity f for instance. So if the application designer wants to use resistive grid templates, the linearity of the grid must be specified; moreover, a higher power consumption or lower speed have to be accepted if high linearities are necessary (cfr. section 2.7) and the application designer should decide if this loss of performance is tolerable.

3.10 ROBUST TEMPLATE DESIGN

The design of templates for CNN's is a nontrivial problem and is often based on intuition or computer simulations. Several algorithms have been proposed to design CNN templates. In [Chu 91] an analytical method for the design of templates is proposed. The desired operation is translated into a set of local rules for the behavior of the cell and its neighbors. This set is translated into a set of linear inequalities for the template elements is derived that ensure the correct operation of the CNN for the given application. A set of inequalities for the template elements can also be derived from the wanted equilibria for the CNN [Zou 90, Vdb 92]. Inequalities that originate from hardware design restrictions can also be included. Inequalities to avoid unwanted equilibria and flow conditions to shape the basin of attractions can further improve the correctness of the learned template [Sei 93]. The template elements are solved by using the simplex algorithm or a relaxation algorithm. Other type of design and learning algorithms have also been developed [Nos 94, and its references] to learn the trajectories of the CNN in time or to design CNN's with genetic algorithms.

Template design for good implementability. For the design algorithms using a set of linear inequalities, the template robustness evaluation method developed in this chapter can be used as a goal function for the optimization of the template elements over the set of solutions of the linear inequalities. In this way, robust functional templates can be designed. For instance, the template elements can be optimized to minimize the required relative accuracy (3.12) for the template circuits for a fixed yield.

116   ANALOG VLSI INTEGRATION OF MASSIVE PARALLEL SYSTEMS

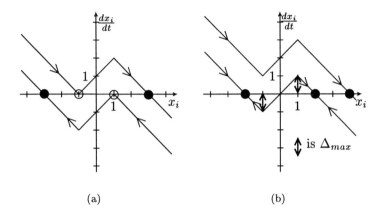

Figure 3.14. (a) The critical dynamic routes for the original holefiller template HOLE; stable equilibria are indicated with a filled circle and unstable with a circle; (b) The critical dynamic routes for the modified template HOLE_MOD; the maximal allowed deviation Δ_{max} for a correct operation is increased from 0 for the original template (a) to 1 for the modified template (b).

This procedure is comparable to the classical design centering approaches [Bra 80, Swi 95]. There first the acceptability region for the parameters is determined. Then a goal function or quality measure is defined and the parameters are redesigned within the acceptability region towards optimal quality. For CNN's the acceptability region is bounded by the inequalities. The quality of a template is determined by its robustness or by the ease of its VLSI realization. This is related to the required accuracy in the circuits which is given by (3.12) or (3.10). Not only a large Δ_{max} is desirable for the template, but also a low value for M_{rel} or M_{abs} which is a property of simple templates. Optimal templates have a high (Δ_{max}/M_{rel}) or (Δ_{max}/M_{abs}) ratio. In [Nac 92], e.g., similar optimization goals are defined for the CNN templates.

Redesign of hole-filler template.   On the left of figure 3.14 the two critical dynamic routes for the classical holefiller template [Mat 90] are drawn. The upper route is for a $k=1$ corresponding to the case of a black input pixel and all but one neighbor's state white. The lower route is for $k=0$ corresponding to a white input pixel and all neighbors' states black or a black input pixel and all neighbors' states white. If the cells all start with an initial state of black and the input image is presented at the inputs, the network will fill up the holes in the image. However for $k=1$ the equilibrium at $x_i = 1$ is unstable and the slightest variation, which is unavoidable in circuits due to the presence of noise, results in an incorrect computation. This template cannot be used in a hardware implementation. If the I-template value is changed from I=-1 to I=0, the cell behavior becomes much more insensitive to non idealities [Kin 94c]. Now a Δ_{max} of 1 is achieved and the higher robustness makes a hardware implementation feasible (see section 3.4). Moreover, the holefiller will now work for black images on white backgrounds and white images on black backgrounds as reported in [Chu 91].

Optimal Connected Component detector. We illustrate the design procedure for a connected component function. An optimal CoCoD template is designed for circuits with a *constant relative error* σ_{rel}. Templates of the form A=[a,a+1,-a], B=0 and I=0 with $a > 0$ have dynamic routes that are isomorf to the standard CoCoD template and all perform the CoCoD function. The Δ_{max} of this template is equal to a.

For a specific network size and wanted yield the necessary $P_{correct}$ can be derived from (3.8) and the maximal $\sigma(\Delta_{route})$ is calculated with (3.6). However from (3.6) and figure 3.3 we can conclude that the larger Δ_{max} of the template is, the larger the allowed $\sigma(\Delta_{route})$ is, and the less accurate circuits are required; the Δ_{max} is thus a good measure for the robustness of the template. On the other hand, the accuracy of the circuits is also dependent on the number of connections and on the value of the template elements (see (3.12)). Consequently a good measure for the optimality of the template is:

$$\frac{\Delta_{max}}{M_{rel}} = \frac{a}{\sqrt{1+a^2+(a+1)^2+(-a)^2}} \quad (3.17)$$

In figure 3.15 equation (3.17) is plotted versus a. The larger the value of a the better the template, but the rate of improvement lowers for larger a values. In practice, the a value will then be limited by a practical limitation like e.g. the limited state swing (see equation (1.3)).

3.11 CONCLUSIONS

In this chapter we have developed the necessary implementation-oriented theory for the VLSI design of cellular neural networks.

Mismatch in the circuit devices result in random variations of the static characteristics of the cells and can cause a faulty cell operation. A procedure to calculate the probability of a faulty cell operation for a given random variation in the template elements has been proposed. Then the probability of a correct network, which corresponds to the yield of the network implementation, is calculated. For large classes of templates the validity of this approach can be

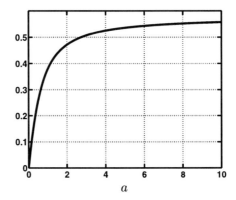

Figure 3.15. The quality of a CoCoD template $[a, a+1, a]$ as function of a for template circuits with a fixed relative error.

proved and for a few templates the validity is checked with extensive Monte-Carlo simulations.

This yield estimation forms the basis for a procedure to derive the accuracy specifications for the different building blocks in the CNN cell implementation. These accuracy specifications play an important role in the optimization of the sizing of the different circuits in chapter 4.

After the random static errors we have treated the effect of random dynamical errors which are related to the mismatch of the time constants of the different cells. For large classes of templates, they introduce no significant deviations and for the propagating templates their impact is much less significant as the impact of the random static errors.

Also significant systematic deviations can occur in very compact circuit implementations of the computational building blocks and as for any deviation, the more relaxed the specifications, the better performance can be attained. The effect of distortion errors or static systematic errors has been studied in detail with a similar method as for the evaluation of the static random errors. We have shown that distortion errors introduce no significant design problems and their effect can be very efficiently modeled and evaluated with the presented methods.

The parasitic poles in the circuit realization of a CNN cell represent systematic dynamic errors. Poles with time constants very close to the cell time constant, cause faulty cell computations and have to be avoided. We have demonstrated that the dominant parasitic time constant should be kept 10x smaller than the cell time constant; this requirement has important implications for the circuit designer as is discussed in chapter 4.

The evaluation methods developed in this chapter, have also been applied to the design of robust templates. The templates are optimized for a high robustness towards the hardware deviations and we have shown that the robustness of several standard templates is improved significantly by changing some of the template parameters. These improvements also result in more relaxed circuit specifications so that a better circuit performance can be obtained.

Throughout this discussion we have pointed out what are the specifications that have to be provided by the CNN application designers for the CNN chip design, and more important, we have indicated the impact of these system-level specifications on the performance of the VLSI implementation.

The methods and results presented in this chapter are aimed at clearly specifying the requirements of the different building blocks in the circuit realization. As such they can be described as implementation-oriented theory for cellular neural networks and are the foundation for the actual VLSI implementation presented in chapter 4.

4 VLSI IMPLEMENTATION OF CELLULAR NEURAL NETWORKS

4.1 INTRODUCTION

The realization of analog parallel computation circuits in VLSI requires the translation of the system model into an equivalent electrical circuit that performs the computations; the circuit structure must be designed such that the differential equation describing the behavior of the circuit is equivalent to the differential equation of the system. Moreover, a high circuit density is desired so that the designer has to come up with compact circuit solutions which consume as less power as possible and have a high speed operation. In order to have access to the high density standard digital CMOS technologies, the analog circuitry should also be fully compatible with these CMOS technologies.

In this chapter the circuit implementation of the basic computational operations is discussed in section 4.2. Especially the realization of programmable scaling of signals and multiplication is difficult to realize with compact CMOS circuits. Two compact multiplier structures based on the operation of MOS transistors in their linear region are introduced in section 4.3.

The input and output of the information into the systems is another important circuit design challenge. The alternatives of transferring the signals from the sensors into the computational system is discussed in detail in section 4.4 and an X-Y addressing topology is proposed for programmable systems.

The chapter then proceeds with the actual implementation of a computational algorithm i.e. a programmable cellular neural network system. A compact VLSI CNN cell architecture is introduced which is fully compatible with a standard digital CMOS technology. On the basis of the specifications and the knowledge of chapter 3 an efficient circuit implementation for each required function can be used and is presented in section 4.5. Building blocks for the automatic biasing and tuning of the analog building blocks from external digital user specifications are also introduced. Two chip realizations are discussed in detail. The feasibility of the realization of a fully programmable CNN is demonstrated with the design of a 4x4 CNN prototype chip (section 4.6). In section 4.7 a large fully programmable CNN system or analog parallel array processor for sensor interfacing applications is presented. For this chip design the mismatch models of chapter 2 and the specifications derived in chapter 3 are applied for the optimization of the circuit sizing towards a high cell density while guaranteeing a high chip yield.

Finally in section 4.8 the performances of the presented CNN chip implementations are evaluated and compared to other analog implementations; furthermore, the performance of the analog systems is compared to the performance of digital implementations.

4.2 ANALOG VLSI IMPLEMENTATION OF COMPUTATION OPERATIONS

4.2.1 Signal representation

In an electronic circuit, a signal can physically be represented under different forms. The following physical quantities are often used in analog VLSI circuits:

Voltage: the difference in electrical potential between two nodes;

Current: the rate of charge flow or current through a wire;

Charge: the amount of charge stored on a capacitor;

Time: the time delay between two pulses;

Frequency: the rate of pulses or the frequency of signals.

Two different approaches for the signal representations are available: continuous time or discrete time signals. The continuous time representations are inherently the fastest when used in circuit applications since the full bandwidth of the technology can be exploited. Discrete time signal representations can however be more robust but require a good settling of the circuit after the processing of each time sample. Therefore, the maximal signal rate is always considerably lower than the full technology bandwidth. This is also why the maximal signal frequencies in digital circuits are always a few orders lower than in continuous time analog circuits. Although many analog circuits like e.g. switched capacitors, use discrete time analog signals, they are not further treated in this work due to their inherent lower speed.

High speed computational circuits rely mainly on voltages and currents to represent the signal properties or the information to be processed, since the other representations can only be used in clocked or discrete-time systems. The choice of the representation depends on the computations that have to be performed on the signal. The optimal circuit implementation of the computations require the signals to be in a certain representation and some computational circuits inherently do a conversion from one representation to another one at the same time as the computation. In the subsequent sections, the different computations and their circuit implementation are discussed.

4.2.2 Addition

The addition of two or many signals is used in many signal processing systems. Massively parallel systems require the addition of a large number of signals which requires a large dynamic range[†] for the output signal. Two voltages are added by connecting them in series. Kirchoff's voltage law states that the total voltage is the sum of the two series-connected voltage signals. In a practical VLSI circuit however, only a small voltage range is available due to the limited supply voltages that can be tolerated on integrated circuits. This makes the addition of voltages un-practical.

Kirchoff's current law states that the sum of currents flowing out of a node is equal to the sum of the currents flowing into the node, due to the conservation of charge[‡]. Adding current signals is thus very easy; as long as the input impedance of the stage processing the sum of currents can be made low enough the maximal output current can be very large. The minimal current that can be used, is limited by the leakage current that is always present in analog circuits. These leakage currents are typically in the order of magnitude of a few pA. If we assume a typical input impedance for the next stage of 100 ohm and allow a maximal voltage drop of 1 Volt, a dynamic range of about 7 decades can be realized for the addition of currents. However, the limited output conductance of practical current sources also introduces errors and restricts this range further. In transistor circuits the input impedance is proportional to $1/g_m$ and

---

[†]In this context the dynamic range of a signal is defined as the ratio of its maximal value over its minimal value.
[‡]Thanks to the conservation of charge, the addition of charge packets is also very easy. Representing a signal as a charge practically requires a time discrete signal and is therefore not further considered.

the output impedance is g_o; the allowed current range is g_m/g_o which is typically about 10 to 100 for a (sub-micron) MOS transistor or only two decades. By applying boosting and cascode circuits techniques e.g. this range is easily extended to about 4 to 5 decades.

4.2.3 Integration

Many algorithms rely on the integration of a signal over time to extract dynamic characteristics of signals [VdS 92] or use dynamical behavior of the system for computations. An integration is just a summation over time of the signal. In practice only a current can be integrated on a capacitor on a VLSI circuit, again thanks to the charge conservation; the voltage across a linear capacitor is:

$$v_C(t) = v_c(0) + 1/C \cdot \int_0^t i_C(\tau) \cdot d\tau \qquad (4.1)$$

The very thin oxides that can be manufactured, make the capacitance per area on a VLSI circuit acceptable. In theory, a voltage could also be integrated by using an inductor, but the inductances of integrated coils are too small for use in computational systems. In very high speed signal processing applications, however, integrated coils can be applied [Cro 96].

When integration over time is used to build a dynamical behavior into the signal processing system, the time constant of the integrator is very important. The time constant of the integrators depends on the resistance that is associated with the voltage signal source. Very small time constants - or very high speeds - are hard to achieve due the required low impedances, which require a high power drain, and the inherent parasitic capacitors in transistors which limit their maximal frequency behavior. On the other hand very large time constants are also difficult to achieve. Only relatively small capacitor values (order Pico-Farad) are available on chip, very large resistances (order Giga-Ohm) are required to obtain large time constants (order seconds) on a reasonable circuit area. Then the integrators become sensitive to leakage currents and have large offsets [Stey91, Kin 92].

4.3 PROGRAMMABLE WEIGHTING OF SIGNALS

A very important operation that is extensively used in massively parallel signal processing systems, is the weighting of signals or multiplication of signals with a constant. In biological neural nets this operation is performed by the synapses and their artificial equivalents are therefore sometimes also called synapses. In a CNN e.g. the signals coming from the neighboring cells are multiplied with the weight factors from the A and B templates; these weights completely determine the operation and behavior of the system. This weighting must be performed on every incoming signal so that the realization of compact multipliers is essential for achieving high cell densities in massively parallel processing systems. The circuit implementation of this multiplication, weighting or scaling is different for fixed scaling factors or programmable scaling factors.

4.3.1 Fixed weighting

With current mirror circuits (see fig. 2.7) a fixed scaling of a current signal is easily attained and a high dynamic range can be obtained. Voltages can be scaled by relying on ratio's of resistors to fix the amplification of a voltage amplifier (see e.g. fig. 2.8) or by using resistive dividers for scaling factors smaller than one but the limited available supply voltage limits the dynamic range for the scaling of voltages considerably.

4.3.2 Multipliers for programmable weights

The implementation of programmable scaling factors requires the realization of a multiplier circuit on chip. Moreover, for many systems a large dynamic range in the weights is required. For the CNN's a range from 1/4 to 4 is typically required or a dynamic range of 16. Mainly four basic principles are available to realize multipliers in an analog CMOS circuit: translinear circuits, square-law circuits, discretely programmable circuits and MOS triode region multipliers.

A   Translinear Circuits.   A first principle that can be applied to realize a multiplication function, is the fact that the logarithm of a product is the sum of the logarithms of the factors:

$$I_1 \cdot I_2 = \exp(\log(I_1) + \log(I_2)) \tag{4.2}$$

This can be used in practical circuits since diffusion currents through an electronic device are exponentially dependent on the voltage drop across the device. These two observation have led to a large class of (non-linear) signal processing circuits, called translinear circuits [Gil 90, See 91].

In a bipolar transistor the collector-emitter current (I_{CE}) is exponentially dependent on the base-emitter voltage (V_{BE}) [Lak 94]:

$$i_{CE} = I_S \cdot \exp(v_{BE}/U_T) \tag{4.3}$$

By connecting several v_{BE} voltages, corresponding to the input currents, in a loop and generating the output current from one of the v_{BE}'s, a multiplier can be constructed. In figure 4.1 a simple translinear multiplier is represented [Gil 90]. By analyzing the translinear loop comprising the four V_{BE} voltages, the

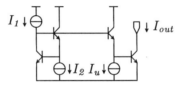

Figure 4.1. Compact bipolar current multiplier based on the translinear circuit principle.

output current is found to be:

$$I_{out} = \frac{I_1 \cdot I_2}{I_u} \tag{4.4}$$

The main advantage of this circuit is the large dynamic range in the signals that can be used; since the exponential relationship between V_{BE} and I_{CE} is valid over 7 orders of magnitude in current, this circuit can handle a large dynamic range in the input signals [Gil 90].

In a CMOS technology, only parasitic bipolar transistors are available [Vit 83]. The *vertical bipolar transistors* use a MOS source as emitter, a well as base and the substrate as collector. The collector terminal is always connected to the negative power supply or the ground potential; this severely restricts the applicability of this transistor. *Lateral bipolar transistors* are built in a well. They have the same structure as a MOS transistor but use a different biasing scheme. All terminals can be freely connected but the substrate is always an extra parasitic collector terminal tied to ground; some of the emitted current does not reach the lateral collector which is a limitation. The ratio between the collector and emitter currents is thus smaller than one due to the substrate current, but remains almost constant for up to 4 orders of current magnitude. By using appropriate configurations translinear circuits can be realized with lateral bipolars [Arr 89]. However, the characteristics and device parameters are not included in the standard set of technology parameters and are also not guaranteed by the chip foundries. This is the most important (practical) limitation for the circuit designer.

The exponential dependence of the drain-source current I_{DS} on the gate-source voltage V_{GS} in the sub-threshold operation region of MOS transistors can be used to make translinear circuits (A.3):

$$i_{DS} = I_{D0} \frac{W}{L} \cdot \exp\left(\frac{v_{GS}}{n \cdot U_T}\right) \tag{4.5}$$

Low current values and large device sizes have to be used for weak-inversion operation. The value of the sub-threshold slope factor n is only constant for a limited biasing range so that only relatively small dynamic ranges are achieved. Moreover, the accuracy of translinear circuits relies on the matching of the specific currents I_{D0} in sub-threshold and thus on the current matching of sub-threshold transistors, which is very poor as is pointed out in chapter 2. I_{D0} depends exponentially on the threshold voltage of the transistors [Tsi 88]. The relative accuracy of the output signal i_{OUT} of the multiplier in figure 4.1, implemented with MOS transistors in weak inversion, is thus given by:

$$\left(\frac{\sigma(I)}{I}\right)_{out} = \exp\left(\frac{2 \cdot \sigma(V_T)}{nU_T}\right) - 1 \qquad (4.6)$$

when n is in first order assumed equal for all transistors. For a 10/10 transistor in the 0.7 μm CMOS technology, this results in a relative accuracy of 11 %. For the same transistor size, a current mirror biased at the edge with weak-inversion, will have an accuracy that is approximately 10 times better.

We can conclude that due to accuracy limitations or practical limitations, translinear circuits cannot be applied in a standard digital CMOS technology to build analog multipliers for the implementation of programmable CNN's.

B Square-Law Circuits. Several classes of computational circuits have been proposed based on the quadratic relationship between the gate-overdrive and the drain-source current for a MOS transistor in saturation [Bul 87]. The translinear circuit technique has also been extended towards MOS circuits [See 91]. The multipliers in these classes are typically realized using the following or a similar mathematical relationship:

$$x \cdot y = 1/4\left((x+y)^2 - (x-y)^2\right) \qquad (4.7)$$

An important limitation in these circuits is the small signal over bias ratio; the biasing signals scale with the *square* of the input signals so that the signal over bias ratio becomes very small when a large range in the inputs is required. This results in high power consumption and high sensitivity to matching errors. These circuit topologies are also sensitive to the bulk effect which can further limit the range of the inputs. Moreover, in sub-micron transistors the current-voltage characteristics only have a very limited operation region, where a quadratic behavior is observed, so that only very small input ranges can be used or high distortion errors occur.

A very common multiplier in bipolar circuits is the Gilbert multiplier [Gil 68], and its operation is based on the linear relationship between the current and the transconductance of a bipolar transistor. In figure 4.2 the circuit schematic for a compact Gilbert-multiplier circuit in CMOS is shown. However, due to

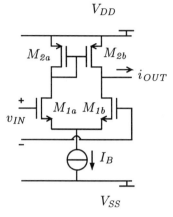

Figure 4.2. Circuit schematic for a compact Gilbert-multiplier in CMOS.

the square law current voltage relationship, the CMOS multiplier has the same disadvantages as any square-law computational block, which we discuss now in detail for this circuit. The linear voltage input range is limited to a fraction of the $(V_{GS} - V_T)$ of the equally sized input transistors M_{1a} and M_{1b}, which is dependent on the bias current I_B and the sizing. The output current is:

$$i_{OUT} \approx \sqrt{\beta_1 I_B} v_{IN} \qquad (4.8)$$

so that the scaling factor can be controlled through the bias current I_B. However the scaling factor or weight is only proportional to the square root of the bias current so that for a dynamic range in the weights DR_w the range in the bias currents will be DR_w^2; for the CNN implementation this results in a bias range of 256 for a weight range of 16. This gives rise to severe design problems. For the large weight factors the power consumption will be very large. In order to limit the consumption, the smallest bias current, corresponding to the smallest weight, must be chosen as small as possible. This results in a very small linear voltage input range for reasonable transistor sizes, so that the sensitivity to mismatch of the input pair is high. Moreover, any error in the current mirror M_{2a}-M_{2b}, results in a relative error in the output current proportional to the bias current and to the weight:

$$\frac{\Delta i_{OUT}}{i_{OUT}} \approx \Delta M \frac{I_B}{i_{OUT}} \propto \text{weight} \qquad (4.9)$$

since the signal over bias ratio i_{OUT}/I_B is proportional to $1/\sqrt{I_B}$ or to $1/\text{weight}$. For large weight values the relative error is thus very large, which is the opposite of the required accuracy dependence on the template values (see fig. 3.7).

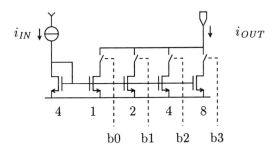

Figure 4.3. A current mirror with discretely programmable mirror factor; the number of unit transistors used for the implementation of each transistor is indicated; the digital control b0-b3 signals determine the mirror factor.

It is clear that for multiplier circuits realized in a sub-micron technology and with a high dynamic range in the weights, circuit topologies based on the operation of transistors in the saturation or quadratic region cannot be used.

C Discrete programmable scaling factor. The above described circuits all implement an analog multiplier function. The connections between cells are programmable over a continuous weight range. However, also discretely programmable connections can be sufficient for some applications. A current mirror with programmable mirror factor is represented in figure 4.3. This circuit can realize a factor from 1/4 to 4 in steps of 1/4. The required accuracy in this implementation is the same as for a multiplier implementation so that in first order the total transistor area will be the same. This circuit requires a lot of extra internal wiring however. Its most important limitation is the large number of global control signals per weight that are necessary to program the weight value: 4 bit-lines for the weight and 1 bit-line for the sign. For an analog multiplier, analog signaling can be used, which requires only two global lines per weight. The number of global wires that have to be routed have to be minimized as much as possible in order to limit the area consumed by routing channels. Therefore analog multipliers result in more compact computing cell implementations.

The weight factor of discretely programmable circuits are only dependent on the size relations between the devices and on local matching between the transistors; basically an A/D conversion is performed locally in every circuit. This makes these circuits more robust. This is an advantage compared to the analog multipliers and the analog signaling; they require extra tuning circuits on chip to generate the correct biasing signals for the multipliers from a digital value supplied by the user. In section 4.5.3 we discuss these tuning circuits for analog multipliers and show that they are compact and can be shared by the cells. If, however, only a very limited programmability is required - e.g. two or four different weight factors - the use of discretely programmable template circuits is a viable approach.

4.3.3 Triode-region MOS analog multiplier circuits

Circuits based on the operation of MOS transistors in the linear region require in first order no linearization or cancelation scheme to obtain a linear circuit for the desired level of accuracy in many parallel analog computational systems.

It is important to note that for the implementation of the weight multiplication a full 4-quadrant multiplier is not really necessary; the weight value remains constant during the computation and only the output changes value during the computation. For the implementation of the sign of the weight, e.g., a switch can be used and a 2-quadrant multiplier circuit is then sufficient. This is more area efficient than implementing a full 4-quadrant multiplier structure.

The current through a MOS transistor, biased in the linear region, is in first order given by (A.13):

$$i_{DS} \approx \beta(v_{GS} - V_T)v_{DS} \qquad (4.10)$$

so that by applying one input voltage signal to the gate of the transistor and the other input voltage signal as the drain-source voltage of the transistor, and by measuring the current through the device a multiplier circuit can be constructed. The transistor is a voltage-controlled-resistor and the multiplication is based on Ohm's Law. The implementation of this principle mainly requires a simple low-impedance current buffer circuit, that allows to apply the voltages and measure the current.

In these topologies, the biases scale *linearly* with the signals so that the signal over bias ratio's remain good even for large dynamic ranges in the inputs; this results in a lower power consumption and a better accuracy of the circuits. These circuits also tend to be more compact. Therefore we use this type of multipliers in the circuit realizations of CNN's.

The CNN template multiplier has two input signals: the A or B template weight signal and respectively the cell output y_i or input signal u_i (cfr. figure 1.3), which we will for convenience both call the input signal of the multiplier in the forthcoming sections. Two alternatives are available for the connection of the weight signal and input signal to the multiplier transistor:

- weight signal to the gate and the input signal to the drain-source;

- or the weight signal to the drain-source and the input signal to the gate.

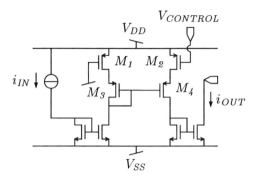

Figure 4.4. Compact programmable current scaling circuit based on the operation of pMOS transistors M_1 and M_2 in their linear region.

Based on these two connection schemes, two triode based MOS linear multipliers are developed in the forthcoming sections.

A   Programmable Current Mirror Circuit.   A *current* scaling circuit has been developed based on triode-biased MOS transistors. Figure 4.4 represents a basic circuit schematic for a programmable current mirror. The transistors M_3 and M_4 are biased in saturation and operate as voltage followers; transistors M_1 and M_2 operate in the linear region as voltage controlled resistors. The weight signal is applied as a voltage $V_{CONTROL}$ at the gate of M_2 and the output current signal is a scaled version of the input current signal. The input-output relationship is given by:

$$\frac{i_{OUT}}{i_{IN}} = \frac{1/g_{ds1} + 1/g_{m3}}{1/g_{ds2} + 1/g_{m4}} \quad (4.11)$$

For ideal voltage followers (M_3 and M_4) with very large g_m, the current ratio is determined by the ratio of the conductances of M_1 and M_2, which are controlled by their gate-overdrive voltages; the current scaling factor is then signal-independent. The follower transistors carry the same current as the conductance transistors so that M_3 and M_4 must be designed much wider than M_1 and M_2 to approach this ideal operation. MOS transistors, however, have only a relatively small g_m for a given bias current[†], so that extra bias current is necessary to increase $g_{m3,4}$ for acceptable transistor sizings. This causes a small signal over bias ratio and thus a high mismatch sensitivity and a high power consumption, as for multipliers based on transistors in saturation.

[†]for a MOS transistor biased at the edge of strong inversion and weak-inversion with a gate-overdrive ($V_{GS} - V_T$) of approximately 200 mV, the maximal g_m/I_{DS} is achieved and is approximately 10; for a bipolar transistor the g_m/I_{CE} ratio is almost independent of biasing and is equal to 40. Therefore a good bipolar transistor would be useful to replace M_3 and M_4 to reduce power consumption or circuit complexity. However, since BiCMOS process developments are typically one generation behind the advanced digital CMOS processes, the overall CMOS implementation will not benefit from an implementation in a BiCMOS process.

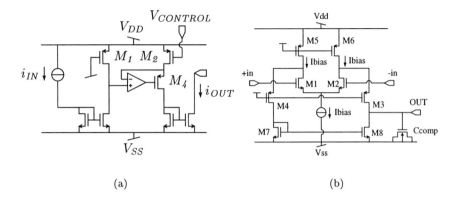

Figure 4.5. (a) Compact programmable current scaling circuit with g_m-boosted cascode transistor. (b) Schematic of the folded cascode boosting OTA; the input voltage range includes V_{DD} for a correct copying of a zero input current.

A better technique is to boost the g_m with the gain of a simple OTA as depicted in figure 4.5(a). The input current generates a voltage across the input transistor M_1; the OTA applies this voltage to the output transistor M_2 so that the input-output current ratio is determined by the conductance ratio's. The cascode transistor M_4 provides a high output conductance and its g_m is boosted by the gain of the OTA. The gate of M_1 is connected to the supply voltage and the weight control voltage $V_{CONTROL}$ is connected to the gate of M_2; using the first order current-voltage relationship for the linear region (A.11), the current scaling factor is:

$$\frac{i_{OUT}}{i_{IN}} \approx \frac{V_{DD} - V_{CONTROL} - V_T}{V_{DD} - V_{SS} - V_T} \qquad (4.12)$$

if the quadratic term in the current voltage relationship can be neglected. For M_1 this is a reasonable assumption since its $(V_{GS} - V_T)$ is large; for M_2 the V_{DS}^2-term becomes larger for higher values of $V_{CONTROL}$ and the maximal distortion error occurs for the small template values. For the large template values, when the current mirror factor is 1, high accuracy is obtained since the quadratic term for both transistors is then equal. The smallest value for $V_{CONTROL}$ is V_{SS} so that the programmable current mirror factor is always smaller than 1: $i_{OUT}/i_{IN} \leq 1$.

For a correct copying of a zero input current, the input voltage range of the OTA has to go up to V_{DD}. This is achieved by using a folded cascode OTA structure with nMOS input transistors as is depicted in figure 4.5(b) [Lak 94, Chapter 6, pp. 587-591].

This multiplier is only a *1-quadrant multiplier* since the weight signal and the input signal must both be positive for a correct operation. As we have already pointed out, the sign of the weight signal is easily implemented by switching the input connection. Since the range of the input current range is limited[†], it can be shifted with a constant bias so that only positive signal currents are obtained. This shifting is implemented easily in the preceding building block and the subtraction of the shift at the output is included in the I-template current.

Circuit design issues. As pointed out in the system level analysis in chapter 3, the cell behavior for a saturated cell output current corresponding to a normalized cell state of ± 1 is especially important. Therefore the accuracy and distortion in the current scaling circuit are investigated for the maximal multiplier input current i_{INmax}. The drain current of transistor M_1 can be modeled by:

$$i_{INmax} = -\beta_1 \cdot (V_{SS} - V_{DD} - V_T) \cdot v_{DSmax} \qquad (4.13)$$

(see (A.13)) since its gate-overdrive voltage is very large; i_{INmax} is determined by the design of preceding building blocks (see section 4.6.2) and the supply voltages are fixed by the technology, so that the choice of v_{DSmax} is the only degree of freedom for sizing transistors M_1 and M_2; v_{DSmax} is optimized to guarantee the accuracy requirements.

Distortion error. For small i_{OUT}/i_{IN} ratio's, the quadratic term in the current-voltage characteristic of M_2 is not negligible and a systematic distortion error appears. Using equation (A.11) we conclude that the output current through M_2 contains a wanted term:

$$i_{OUT} = \beta_2 (V_{DD} - V_{CONTROL} - V_T) \cdot v_{DSmax} \qquad (4.14)$$

and an unwanted distortion term $\Delta i_{OUT} = \beta_2 (v_{DSmax}^2)/2$; the relative distortion error is then:

$$\left(\frac{\Delta i_{OUT}}{i_{OUT}}\right)_{disto} = \frac{v_{DSmax}}{2 \cdot (V_{DD} - V_{CONTROL} - V_T)} \qquad (4.15)$$

For a range in the weights DR_w the maximal error occurs for the smallest weight factor and thus the smallest $(V_{DD} - V_{CONTROL} - V_T)$ equal to $(V_{DD} - V_{SS} - V_T)/DR_w$, and is given by:

$$\left(\frac{\Delta i_{OUT}}{i_{OUT}}\right)_{disto\ max} = \frac{DR_w \cdot v_{DSmax}}{2 \cdot (V_{DD} - V_{SS} - V_T)} \qquad (4.16)$$

[†]both the output signal and input signal are bounded to [-1,+1] in a CNN.

Large distortion errors result in a non-constant mirror factor and in an asymmetric behavior of the multiplier for small and large input signals, which corresponds to an asymmetric cell characteristic. The DC shift in the output signal is then also not constant, which introduces biasing problems. Therefore the distortion error must be kept as small as possible and thus a small v_{DSmax} must be designed for.

Accuracy. In a practical inplementation the OTA has a random input offset voltage $\sigma(V_{os})$ which introduces a difference in the v_{DS} voltages across M_1 and M_2. Together with the mismatch in the β and V_T of M_1 and M_2, this results in a random error on the output signal. The offsets and mismatches are independent random variables with a normal distribution, so that the random error on the output current is derived from (4.14) using (2.12); the maximal random error in the output signal occurs for the smallest weight factor and is given by:

$$\left(\frac{\sigma(\Delta i_{OUT})}{i_{OUT}}\right)^2_{ran.\ max} = \left(\frac{\sigma(V_{os})}{v_{DSmax}}\right)^2 + \left(\frac{\sigma(\beta)}{\beta}\right)^2 + \left(\frac{\sigma(V_T) \cdot DR_w}{V_{SS} - V_{DD} - V_{T0}}\right)^2 \quad (4.17)$$

When v_{DSmax} is chosen too small, large random errors in the weight values will occur.

A trade-off between the distortion specification and accuracy specification can be made through the choice of the v_{DSmax} voltage. In section 4.6.2 we show that a value of 200 mV is a good compromise for the circuit implementation of the programmable current mirror in a 2.4 μm CMOS technology. The accuracy of the multiplier is then mainly dependent on the offset of the boosting OTA; the accuracy is thus independent of the scaling factor so that for a CNN chip application the accuracy behavior of the template circuit can be modeled with a constant absolute error (see section 3.4). The distortion is small enough to limit the cell asymmetry and the sensitivity to mismatch errors is small enough for reasonable transistor sizes.

The g_m-boosting OTA is used in a feedback-configuration inside the multiplier. To obtain a good settling and a stable feedback-loop the phase- and gain-margin must be made large enough (for a detailed design plan of the OTA see [Lak 94, chapter 6, pp. 587-591]). The GBW of the OTA must be made large enough to reduce the parasitic time constant it introduces, to a value of about 10x smaller as the cell time constant (see section 3.7).

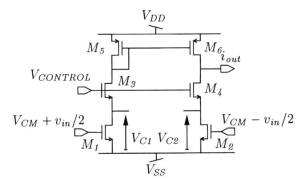

Figure 4.6. A compact 6 transistor grounded transconductance multiplier based on the operation of the input transistors M_1 and M_2 in the linear region.

When a value for i_{INmax} is given, which is determined by the design of the preceding building blocks, the transistors of the current scaling circuit can be sized by inverting equation (4.13) for β_1 using the optimal value for v_{DSmax} derived in this section. The detailed sizing of this circuit is further discussed in section 4.6.2.

B Linear Grounded Transconductance Multiplier. An important limitation in the programmable current mirror is the second-order distortion that is introduced by the v_{DS}^2 term in the current-voltage relationship; since the input signal is represented as the drain-source voltage, this distortion causes asymmetry in the cell behavior for positive and negative cell output or input signals.

The voltage signal at the gate terminal of the triode biased transistor causes in first order no distortion components so that, if the input signal is connected to the gate and the weight signal is connected to the drain-source, a symmetrical cell characteristic is obtained. The transconductance multiplier circuit schematic of figure 4.6 uses this principle. The identical input transistors M_1 and M_2 are biased in the triode region by the cascode transistors M_3 and M_4 that are biased in saturation. The input transistors (M_1 and M_2) are long and the cascode transistors (M_3 and M_4) are wide so that the drain voltages $V_{DS1} = V_{C1}$ and $V_{DS2} = V_{C2}$ remain almost equal and constant: $V_{C1} \approx V_{C2} \approx V_{CONTROL} - V_{GS3,4}$. The current through an input transistor is then calculated from (A.11) as:

$$i_{DS} = \beta_{1,2} \left((v_{GS} - V_T)(V_C) - V_C^2 \right) \qquad (4.18)$$

and contains only a distortion component related to the weight control signal V_C.

By using a differential input signal $V_{CM} \pm v_{in}/2$ and by taking the differential output current i_{OUT} with the current mirror M_5-M_6, the output current is (for an ideal current mirror):

$$i_{out} = i_{M1} - i_{M2} = \beta_{1,2} v_{in} V_{C1,2} \qquad (4.19)$$

and the distortion component related to the weight control signal is suppressed in first order. The template weight is controlled through the V_C and thus the $V_{CONTROL}$; weight tuning circuits are used for the generation of the $V_{CONTROL}$ voltage (see section 4.5.3). The differential input signal is a bipolar signal whereas the weight signal can only be positive so that a *two-quadrant* multiplier circuit is obtained. With an analog multiplexer at the input the sign of the template weight is made switch-able. In section 4.5.1 we show that the differential driving voltages for the multipliers are easily derived from the state-to-output converter or the input buffer in the cell circuit schematic.

Circuit design issues.

Biasing. To obtain a correct operation of the multiplier all transistors must be biased in their correct operation region:

Input transistors: The input transistors M_1 and M_2 must be biased in the linear region; this results in the following condition for the maximal amplitude of the differential input signal V_{inmax} and the maximal weight tune voltage V_{Cmax}:

$$V_{inmax} + V_{Cmax} + V_{T3,4} \leq (V_{DD} - V_{SS}) \qquad (4.20)$$

These limitations are illustrated in figure 4.7.

Cascode and Current Mirror transistors: The cascode transistor M_3 must be operating in the saturation region so that:

$$V_{DD} - V_{SS} - V_{GS5} - V_C \geq (V_{GS} - V_T)_3 \qquad (4.21)$$

which is puts restrictions on the sizing of the current mirror and the cascode transistors. In order to keep the M_4 and M_6 in saturation only a very limited peak-to-peak swing $V_{outppMax}$ at the (current) output can be tolerated:

$$V_{outppMax} \leq V_{DD} - V_{SS} - V_C - (V_{GS} - V_T)_{3,4} - (V_{GS} - V_T)_{5,6} \qquad (4.22)$$

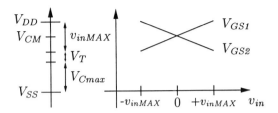

Figure 4.7. The limits on the signal swings and the necessary differential input signals for the transconductance multiplier of figure 4.6.

Many alternative output circuit topologies to the simple current mirror can be thought of to improve the output swing and the output impedance. However, they all increase the circuit complexity and the number of devices and increase the power consumption. Therefore this limitation is circumvented much more efficiently at the cell circuit system level by connecting the output to a single low input impedance current buffer at the cell input (see section 4.5.1).

By using a differential topology the signal/bias ratio decreases; basically each of the input transistors has an extra bias component proportional to $V_C{}^2$. The lower the signal/bias ratio, the more sensitive the multiplier is to errors in the current mirror circuit M_5-M_6. For the current mirror the signal/bias ratio is:

$$\frac{i_{signal}}{I_{BIAS}} = \frac{v_{in}}{2(V_{CM} - V_{T1}) - V_C} \qquad (4.23)$$

which is minimal for small weights and maximal for large weights. The current mirror mismatch sensitivity for the larger weights is thus smaller as for the small weights as is required by the specifications (see fig. 3.7).

Distortion. Equation (4.19) is only valid if both input transistors are biased with the same V_{DS} voltage; this can only be achieved if the current buffer function implemented by the cascode transistors has an zero input impedance or if $g_{m3,4}$ are infinite. For MOS devices the transconductance is rather limited. Since the cascode devices carry the same current as the input devices, the cascode input impedance can only be lowered by using wide cascode transistors and long input transistors. The larger the $\beta_{3,4}/\beta_{1,2}$ ratio, the better the V_{DS} voltage remains constant.

For large input voltages and output currents, the effective V_C value decreases so that the weight value decreases. However, this distortion is symmetric for negative and positive currents. Moreover, this distortion is compensated by the weight tuning circuits (see section 4.5.3) so that for the CNN cell implementation no extra measures have to be taken to eliminate this distortion.

Common-mode suppression. The output current of the multiplier is a bipolar signal; the common-mode components in the currents of the input devices are suppressed by the current mirror. The common mode rejection ratio [Lak 94] for the input signal is thus mainly determined by the matching of the current mirror factor M of current mirror M_5-M_6, of the input devices and of the weight tune voltage V_C:

$$CMRR = \left(\frac{\Delta \beta}{\beta}\right)_{1,2} + \frac{\Delta V_{C1,2}}{V_C} - \Delta M \qquad (4.24)$$

For typical standard deviations of the different components a $CMRR$ of over 35 dB is obtained, which is more than sufficient for the CNN application.

Accuracy. For a current mirror factor M of M_5-M_6, the output current is given by $i_{out} = M \cdot i_{M1} - i_{M2}$; from this equation and (4.18) we conclude that the accuracy of the output current is determined by:

$\Delta V_{T1,2}$: the mismatch of the threshold voltage of the input transistors;

$\left(\frac{\Delta \beta}{\beta}\right)_{1,2}$: the mismatch of the current factors of the input transistors;

$\Delta V_{C1,2}$: the mismatch of the drain-source voltages of the input transistors and thus the mismatch of the gate-source voltages of the cascode transistors;

ΔM: ideally the current mirror factor M is 1, however in practice the current mirror error ΔM in M_5-M_6 occurs.

and is calculated by using (2.11) as:

$$\left(\frac{\Delta I_{out}}{I_{out}}\right) \approx \frac{\Delta M}{2} + \frac{(\Delta V_{T1,2} + \Delta V_{C1,2})}{v_{in}}$$
$$+ \left(\left(\frac{\Delta \beta}{\beta}\right)_{1,2} - \Delta M + \frac{\Delta V_{C1,2}}{V_C}\right) \cdot \frac{I_{DC}}{i_{out}} \qquad (4.25)$$

so that by using (2.12) the standard deviation in the output current is given by:

$$\left(\frac{\sigma(I)}{I}\right)^2_{out} = \left(\frac{\sigma(V_{T1,2})}{v_{in}}\right)^2 + \left(\frac{\sigma(\beta)}{\beta}\right)^2_{1,2} \left(\frac{V_{CM} - V_{T1,2} - V_C/2}{v_{in}}\right)^2$$
$$+ \sigma^2(M) \left(\frac{1}{2} + \frac{V_{CM} - V_{T1,2} - V_C/2}{v_{in}}\right)^2$$
$$+ \sigma^2(V_{C1,2}) \left(\frac{1}{v_{in}} + \frac{V_{CM} - V_{T1,2} - V_C/2}{v_{in}V_C}\right)^2 \quad (4.26)$$

The sensitivity of the multiplier to mismatches in the input transistors mismatches decreases for higher input voltage swings; the signal/bias ratio also improves so that the sensitivity to current mirror errors also decreases. But, if the input swing is made too large the range in the tune voltages V_C will become very small (see equation (4.20)), so that the sensitivity to the errors in the tune voltages will increase. However, the total effect of increasing the input voltage swing is much larger, so that an optimal choice of the $V_{inPPmax}$ is typically larger than V_{Cmax} (see section 4.7.3). This implies that the accuracy is only weakly dependent on the template value; for larger template weight the accuracy improves slightly. For the application of this multiplier to the VLSI design of CNN chips, the accuracy of this template multiplier can thus be modeled with a constant absolute error independent of the template value (see section 3.4).

The combination of the biasing, distortion and accuracy considerations enable an optimization of the circuit sizing to attain a given performance. In section 4.7.3 this optimization is discussed in detail along with the total cell circuit optimization and sizing.

C   Other Circuit Implementations. Many other circuits exploit the linear behavior of a transistor biased in the triode region. MOSFET-C filters use triode-biased MOS transistors as variable resistors to implement integrators. The balanced input signal (V_x) is applied at the drain and the tune voltage (V_C) at the gate as illustrated in figure 4.8(a)[Tsi 86]. When a fully balanced implementation is used, the second-order distortion components of the input signal are suppressed. Consequently complex fully balanced opamps[†] are used to build a low input impedance current measuring circuit. This technique can achieve very high performances, but for analog parallel processing algorithms

---

[†]a balanced opamp is a differential opamp with the common-mode output voltage at ground.

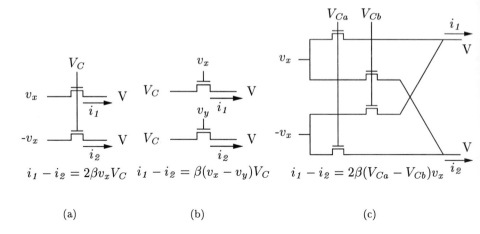

Figure 4.8. Circuit alternatives to compute a multiplication with MOS transistors biased in the linear region: (a) configuration used in MOSFET-C filters; (b) configuration used in our grounded transconductance multiplier circuit; (c) improved configuration used in MOSFET-C filters combining (a) and (b).

this level of performance is not necessary; therefore simpler and lower performance implementations as presented above are preferable.

The principle illustrated in figure 4.8(b) [Cza 86, Son 86, Kha 89] has been used in the above presented multiplier; the input signal $(v_x\text{-}v_y)$ is applied at the gates and the tune signal (V_C) is applied at the drain. In [Kha 89] a computation circuit is proposed, based on the same triode-biased MOS transistors interconnection. However, again complete opamps are used to obtain a highly linear circuit but which is too complex and consumes too much power.

Finally a combination of both previous basic configurations has been proposed in [Cza 86, Son 86] and is represented in figure 4.8(c). It further improves the performance of MOSFET-C filters by making them insensitive to distortion caused by the substrate voltage modulation and also less sensitive for substrate noise. However, these high performances are not required in CNN implementations.

Instead of using full balanced opamps to obtain very low impedance inputs or virtual ground nodes, the g_m of the cascode transistors can also be boosted with an active amplifier; in [Cob 94] e.g. a boosted and four-quadrant version of the multiplier proposed in this section is described. The current buffer circuit that is discussed in [Kin 96b] could also be used to improve the linearity of the multiplier for the weight control voltage input.

4.3.4 Summary

We can conclude that in order to realize a large dynamic range for the programmable template weight factor, scaling or multiplier circuits based on the operation of transistors in the linear region are the best choice. They are compact, are least sensitive to the influence of matching errors and do not have excessive bias signals or large power consumption for large weights.

By combining the knowledge of the system-level extraction of specifications for the building blocks and using tuning techniques to compensate systematic errors, very simple circuit topologies for the template multiplier circuits are developed.

4.4 INPUT-OUTPUT (I/O) CIRCUITS

This section discusses the different hardware alternatives for the input/output circuits of cellular neural networks [Kin 94d]. CNN's are mainly applied in 2D sensor processing applications, like e.g. image processing. The overall speed of the system is also dependent on the speed of the transfer of the outputs of the sensor to the inputs of the CNN. CMOS-compatible sensors can be included within the cell hardware so that no input transfer is necessary; an interface to output the processed result must still be provided. Or a separate sensor can be used, but then a fast and compact input circuit is required; the accuracy of the input interface must be higher than of the output interface since it has to transfer the analog input information, whereas for most of the templates, the outputs or results are black and white and thus very robust.

4.4.1 Specifications for the I/O hardware

Real-time image processing applications require a frame-rate of 25 images per sec so that 40 ms are available to capture, process and output an image[†]; this speed requirement is also representative for other 2D sensor signal processing applications.

Let's assume the following time schedule: 10 ms for the transfer of the image from the sensor into the CNN; 20 ms for the processing; and 10 ms to output the processed image. The integration time of the image sensor is then 30 ms. Other time schedules can e.g. allocate more time for the processing and less for the I/O so that more complex algorithms can be executed. It is clear that the speed of the I/O circuits can be exchanged for the speed of the computation cells to reach a given number of instructions or templates that can be executed per frame.

[†]this is a worst case specification since in some systems a pipelining architecture can be used.

The accuracy used in image processing is commonly between 6 to 8 bits or 1.6 to 0.4 %. This specification limit the maximal allowed signal degradation introduced by the input structures.

4.4.2 Sensors in the cells

Sensors that are compatible to a standard CMOS technology, can be included into the cell's hardware directly. In this way no input transfer is necessary. Mechanical sensors can be made using surface- or bulk-micro-machining techniques. These sensors tend to be rather large; they are mostly fabricated as separate devices since they would decrease the cell density too much. In standard CMOS technologies several solid-state light sensitive devices can be fabricated: photo-diodes, photo-transistors and charge coupled device structures. These devices are used to build CMOS imagers [Ack 96]. The combination of these devices with analog circuitry has resulted in different types of smart image sensors for fixed applications like edge detection, range finding, and spot intensity or position measurements [Mea 89, Kob 90, Gru 91, Sta 91, Yu 92, Che 95, Veni96, Arr 96, Esp 94b]. In these chips the cells consist of a photosensitive device and the analog computation hardware. For a good resolution and sensitivity of the sensor the pitch of the sensors[†] must be small and the fill factor[‡] must be large.

The feasibility of this approach depends on the degree of programmability of the computation array. For a highly programmable array the larger computation cell size results in lower sensitivity and resolution performance; to improve the sensor performance they have to be implemented outside the array. For a fixed function computation array, the cell size is much smaller and the density is larger so that the good sensors can be integrated in the cell array [Esp 94b]. Even with the sensors included in the cells, extra hardware has still to be added to output the processed image.

4.4.3 Separate Sensors and Interface topologies

The image sensing and processing circuitry should be separated for highly programmable analog computational circuits so that both systems can be optimized independently towards high performance. The CMOS compatible sensing devices can still be integrated on the same die so that a single chip solution is obtained. An area efficient, accurate and fast interface is necessary for the transfer of the raw sensor signal into the computation cells and to communicate the processed sensor signal to the output; the same type of architecture is used for external sensors.

[†]the pitch of the sensors is the distance between the centers of the sensors.
[‡]the fill factor is the ratio of the light sensitive area to the non sensitive area.

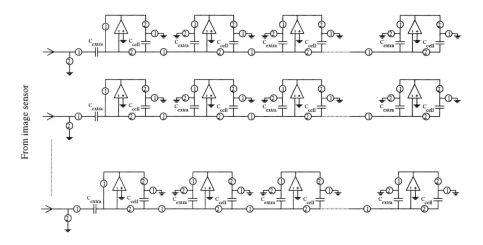

Figure 4.9. A switched capacitor analog shift register with offset compensation; in every cell an extra hold capacitor C_{extra} and opamp has to be added.

4.4.3.1 Analog shift register. In a frame transfer charge coupled device the image is captured during the integration period and then transferred in parallel to the storage register; from the storage area the charges are read out sequentially and converted from a charge signal into a voltage signal [Flo 85]. A charge coupled device array is basically a shift register so that by adding an analog shift register per row in the CNN, the image can be transferred directly into the CNN and stored; the CNN then processes the image and outputs the result.

In figure 4.9 an analog shift register schematic is shown with offset cancelation [Gre 86]. In table 4.1 the required clock frequencies for the double phase clock are given for two image sizes. To initialize one capacitor in the cell with the shift register an extra temporary hold capacitor has to be introduced and an opamp has to be included. In table 4.2 the opamp specifications for a precision of 6 bits are summarized. Three main error sources exist. The *random error source* V_{OS} is the remaining offset for each transfert, mainly due to clock feed-through. The error ($\sigma(V_{error})$) at the end of the line - where we still want a precision of 6 bits - is a contribution to the error proportional to the $\sigma(V_{os})$ per cell and the square root of number of cells N shifted through:

$$\sigma(V_{error}) = \sqrt{N} \cdot \sigma(V_{OS}) \tag{4.27}$$

Table 4.1. Clock frequency for different I/O schemes for a fixed transfer time of 10 ms.

	128×128	512×512
SHIFT REGISTER	25 kHz	100 kHz
X-Y ADDRESSING parallel	12.5 kHz	50 kHz
serial	1.6 MHz	26 MHz

Table 4.2. Shift register opamp specifications for a precision of 6 bits, a signal of 1 V_{pp} and the clock frequencies of table 4.1.

	128×128	512×512
RANDOM ERROR V_{OS}	0.5 mV	0.23 mV
SYSTEMATIC ERRORS DC Gain	88 dB	100 dB
GBW	161 kHz	732 kHz

Two systematic error sources exist: the relative static error (ϵ_s) due to the finite opamp DC gain, puts a lower limit on the required gain:

$$\epsilon_s = V_{es}/V_{in} = 1/A_{DC} \tag{4.28}$$

The relative dynamic error (ϵ_d) due to the limited bandwidth of the opamp, puts a lower limit on its Gain-Bandwidth product (GBW):

$$\epsilon_d = V_{ed}/V_{in} = \exp\left(\frac{-\pi \text{GBW}}{f_{clock}}\right) \tag{4.29}$$

The total contribution of these systematic errors for each transfer at the end of the line V_{error} is the accumulation of the error in each step; it is proportional to the number of cells N:

$$V_{error}/V_{in} = (1 + \epsilon_s + \epsilon_d)^N \approx (1 + N \cdot (\epsilon_s + \epsilon_d)) \tag{4.30}$$

From table 4.2 it is clear that the offset specification is very hard to achieve since a typical order of magnitude for clock feed-through is 10 mV/pF [VPg

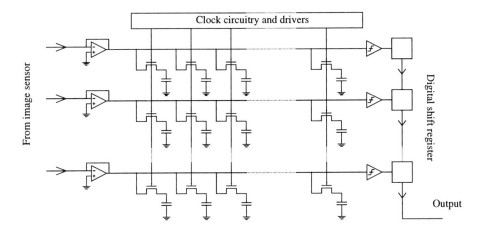

Figure 4.10. An X-Y addressing I/O scheme; a switch per cell and a buffer and a comparator per row are added; the digital register is used to obtain a serial output.

86]. With proper clock feed-through cancelling techniques, it can be reduced down-to 1 mV/pF at the cost of extra clock phases and circuitry. To achieve the specification a large cell capacitor is required, but this increases the time constants of the cells. The opamp gain and frequency specifications are acceptable and could be exchanged for easier offset specifications but this results in a more complex and larger opamp. Clearly, this I/O scheme is area consuming since two (large) capacitors and an extra high performance opamp have to be included in each cell.

4.4.3.2 X-Y addressing scheme. In figure 4.10 a parallel X-Y addressing scheme is drawn. The capacitors of the cells in one column are all initialized in one clock cycle. The serial version of I/O scheme is widely used in applications like RAM or X-Y addressable photo-diode image sensors [Ack 96]. In the CNN application it can be made fully compatible with the different types of image sensors.

The specifications of the different subblocks in the system are summarized in table 4.3. Two limitations exist. The buffer has a high capacitive load of the long line and the high number of drain bulk capacitors of the switches. However only one buffer per row is necessary so that its area and power consumption can be higher. The buffer specifications are derived from similar considerations as the opamps in the shift register; the required performance is no problem for a CMOS realization.

Table 4.3. Row buffer specifications for X-Y addressing for a precision of 6 bits, a signal of $1V_{pp}$ and the clock frequencies of table 4.1 ; the load capacitors are calculated for 0.7 μm technology.

	128 × 128	512 × 512
CAPACITIVE LOAD (line & switches)	$2pF$	$10pF$
BUFFER		
V_{OS}	$15mV$	$15mV$
DC Gain	$42dB$	$42dB$
GBW	$20kHz$	$80kHz$

The values put on the capacitors in the first column have to wait until the last column is initialized before they can be processed by the CNN. The degradation due to the leakage current should be small enough to guarantee the accuracy. For a leakage current of 10 $fA/\mu m^2$, a drain area of the switch of $5 \times 5 \mu m^2$ and a cell capacitor of 1 pF[†], the hold time is 160 ms for an accuracy of 8 bits and a signal of 1 V, whereas the input of the whole image takes 10 ms. Clearly the X-Y addressing sub-blocks have easier specifications, since the errors are not accumulated. The extra hardware to be added is limited to a switch per cell and a buffer per row.

With this scheme it is also possible to provide *serial* output of the image in 10 ms. The clock frequency for outputting must then be increased by a factor equal to the image size (see table 4.1). The output image is black and white so that a digital comparator at the end of the line and a digital shift register can be used. This requires digital circuitry running at $26MHz$ which is no problem for sub-micron technologies.

4.4.4 Conclusions

We can conclude that on fully programmable analog computation chips, the sensors are better physically separated from the computation electronics in order to improve the sensor's performance. Efficient X-Y addressing based I/O schemes can be included in the computation electronics to interface with 2D array sensors that can be located on-chip or off-chip. With the present sub-

---

[†]in a 0.7 μm technology this switch and hold capacitor combination allow a clock frequency of over 200 MHz.

micron technologies these interfaces achieve very high speeds at the required accuracy levels and are very compact.

4.5 VLSI CELL ARCHITECTURE FOR CNN-CELL

In this section an architecture for the circuit implementation of a CNN cell with analog VLSI circuits is presented. The main objectives in deriving a circuit architecture are:

- compatibility with a *standard digital CMOS technology*;
- compact and elegant circuit solutions;
- and easy external interfacing for the user.

Analog parallel computation applications require a large number of cells to be integrated on a single chip die, so the main objective for the VLSI implementation of the cell is its area compactness and simplicity of implementation. Moreover, to reduce cost and to have access to the finest line-widths available, the analog circuits have to be compatible with a standard digital CMOS process. Analog extensions for CMOS technologies to implement very linear capacitors or high-ohmic linear resistors require extra mask sets; they increase the processing cost and are not always available. Moreover, these devices are not really required for the implementation of analog parallel computation applications. Thanks to the knowledge of chapter 3 we can evaluate the impact of non-linearities for example, and still use very simple circuit solutions to replace the linear components with transistor structures.

To achieve a compact cell implementation the most compact implementation for the different functions must be used; in a CNN cell above all a compact multiplier is necessary whose implementation is discussed in section 4.3.2.

To make the application of analog parallel computation system easy, all biasing circuits are included on-chip. Moreover, to simplify the interfacing to the system, a digital interface is provided, which derives on-chip the correct control signals for the cells.

4.5.1 Block diagram

A Signal Representation. The transformation of the mathematical cell model of figure 1.3 into an electronic circuit, suited for analog VLSI implementation, requires the representation of the signals as physical quantities. In figure 1.4 a general schematic for an electronic cell implementation is given. The mapping of the signals onto physical signals is mainly determined by the requirements of the circuit implementation of the mathematical operations (see

148 ANALOG VLSI INTEGRATION OF MASSIVE PARALLEL SYSTEMS

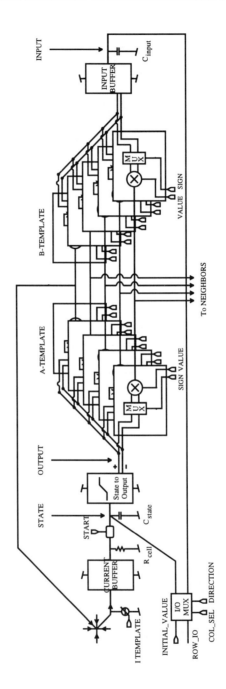

Figure 4.11. VLSI compatible circuit implementation for a CNN cell.

section 4.2.1). In the theoretical analysis of CNN's all signals are normalized to unity. For a VLSI implementation a scaling of the signals to voltages and currents must be performed. For the current signals a unit current I_{UNIT} and for the voltage signals a unit voltage V_{UNIT} is defined. The unit interval [-1,1] is then mapped to [-V_{UNIT},V_{UNIT}] in voltage and [-I_{UNIT},I_{UNIT}] in current. Figure 4.11 presents the circuit architecture for a fully programmable analog CNN cell suited for VLSI implementation.

On a VLSI chip integral computation can practically only be performed by integrating a current on a capacitor. The *state* x_i of the cell is represented as the voltage across the state capacitor C_{state}. The loss for the integrator is realized with a cell resistor R_{cell} in parallel with the state capacitor C_{state}.

The *communication signals* between the neighbors and the feedback signals have to be summed with high precision; therefore they are represented as currents. The *I-template* signal is also represented as a current signal and summed with the communications signals.

The *input* u_i of the cell remains constant during the cell evolution. It is memorized as a voltage across a capacitor C_{input} used as a dynamic analog memory; the retention time of this memory is sufficient, since the computation speed of the network is high. The input node cannot be loaded so that an input-buffer circuit is included, which also converts the single-ended input signal in the correct format for the analog B-template weight multipliers.

Depending on the type of analog multiplier, its input signals must be a current or a voltage. The state-to-output circuit and the input buffer circuit deliver their output signal in the correct form; these blocks are thus V/I or V/V converters.

B Standard CMOS circuit implementations.

Capacitors. In a most standard digital CMOS processes no analog capacitors or high-ohmic resistors are available; therefore the *state C_{state} and input capacitors C_{input}* are implemented by using the gate capacitor of a transistor, which is biased in strong inversion. This capacitor has enough linearity for the CNN application.

Cell resistor. The *cell resistor* R_{cell} is built with a nMOS transistor and pMOS transistor in parallel biased in the linear region as shown in figure 4.12(a). This compact structure is a grounded resistor for all state voltages and has a constant resistance in the unit range. This guarantees a correct operation, since the exact evolution of the state is only critical in the unit-range. The resistance

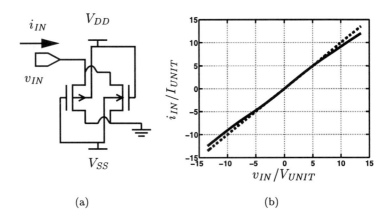

Figure 4.12. (a) Grounded cell resistor structure implemented with a parallel connection of an nMOS and pMOS transistor biased in the linear region. (b) Simulation of the resistor structure of (a) implemented in a 2.4 μm technology with $V_{UNIT} = 300\ mV$ and $I_{UNIT} = 1.25\ \mu A$; the current-voltage characteristic of (a) is represented by (–) and the current-voltage characteristic of an ideal 240 $k\Omega$ resistor is (- -).

R_{cell} around ground is calculated using (A.13) as:

$$1/R_{cell} \approx \beta_n \cdot (V_{DD} - V_{Tn}) + \beta_p \cdot (|V_{SS}| - |V_{Tp}|) \tag{4.31}$$

In figure 4.12(b) the simulated I-V characteristic of the cell resistor circuit implemented in a 2.4 μm technology is compared to the characteristic of an ideal resistor. For transistor sizes $\left(\frac{W}{L}\right)_n = 2.4/156$ and $\left(\frac{W}{L}\right)_p = 2.4/42.5$, and for supply voltages $V_{DD} = 5$ and $V_{SS} = -5$, the predicted resistance value is $\approx 240\ k\Omega$ using (4.31) and the process data of the 2.4 μm technology in appendix A taking into account the mobility reduction [Lak 94] ($K_{Pn} = 43\ \mu A/V^2$ and $K_{Pp} = 15\ \mu A/V^2$) and the increase of the threshold voltages due to the bulk-effect ($V_{Tn} = 1.65\ V$ and $V_{Tp} = 2.4\ V$).

The resistance remains constant up to ±5 times the unit voltage and decreases slightly for very large state voltages. However, this has no influence on the correct operation of the CNN cell (see section 3.6) and the obtained linearity is more than sufficient.

State-to-output buffer. The *state-to-output buffer* has two functions. It implements the non-linear transformation of the state variable into the output variable and it converts the single-ended state voltage into a suitable form to drive the multiplier circuit. The tanh-like non-linear output current of a differential pair circuit approximates well the ideal piece-wise linear characteristic of

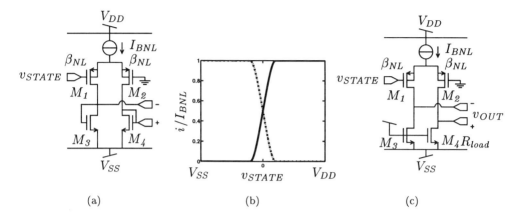

Figure 4.13. Compact CMOS implementation of the non-linear state-to-output converter with a current output (a) or a voltage output (c); (b) the distribution of the bias current over the two output branches as a function of the state voltage for the circuit in (a): (-.-) i_{M3} and (–) i_{M4}.

the non-linear V/I. In figure 4.13(a) a state-to-output converter with current output is represented; the current outputs are connected through a multiplexer to the programmable current mirror of figure 4.5(a). The current in the output branches as a function of the state voltage is represented in figure 4.13(b). When a differential pair is loaded with resistors, implemented as pMOS transistors biased in the linear region, as in figure 4.13(c) a differential voltage output is obtained that can drive the analog transconductance multiplier of figure 4.6.

Input buffer. The *input buffer* has to buffer the input capacitor and has to convert the single-ended input voltage into a current or into a differential voltage depending on the type of multiplier used for the B-template weights. A differential pair can perform both functions and by proper sizing the transfer characteristic of the differential pair can be linearized over the unit range. In figure 4.14(a) a circuit schematic for the input buffer with current output is represented. In the circuit of figure 4.14(b) the differential pair is loaded with resistors to obtain a voltage output; at the same time, a 'Krummenacher-degeneration' is used to improve the linearity without increasing the bias current or the power consumption [Kru 88].

Current buffer. At the input of the cell a *current buffer* is implemented. It buffers the current input from the large voltage swing at the state node during the computation. If the multiplier circuits, that are tied to the current input

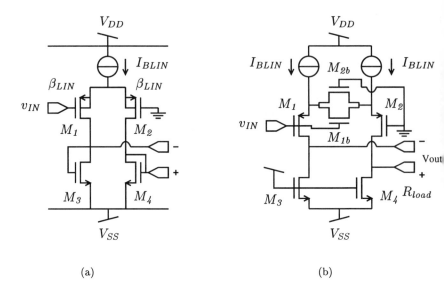

Figure 4.14. Compact CMOS implementation of the input buffer with a current output (a) or a voltage output (b); the circuit in (b) also uses a 'Krummenacher' degeneration to improve the linearity.

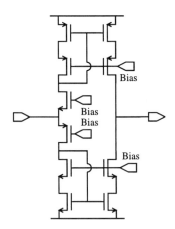

Figure 4.15. Compact CMOS implementation of a current buffer function.

of a cell have a high output impedance, the current buffer is not necessary. However, in many cases it is more efficient in terms of power and area consumption to implement only one current buffer in the cell, than to add extra circuitry to improve the output impedance of the multipliers. In figure 4.15 a CMOS schematic for a current buffer or operational current amplifier (OCA) is represented (for a detailed design plan of this circuit see [Lak 94, chapter 6]). Its input impedance is determined by the transconductance of the input devices, which are made wide. The output impedance is improved by adding cascode transistors; to improve the allowed output swing, the cascode devices are biased with a low gate voltage; to reduce the systematic error in the current mirrors, the drain-source voltages of the mirror transistors are kept equal by the cascode transistors. Thanks to the class AB operation of this circuit, it can be biased with a small bias current to reduce static power consumption; this biasing is controlled with an extra replica biasing circuit that is implemented once on the chip for all cells.

Control and I/O. In order to start and stop the analog computations by the cell a START switch is included between the output of the current buffer loaded by the cell resistor and the state capacitor and input of the state-to-output converter (see fig. 4.11). When this switch is open, the integrator does not receive any input signal and the cell remains in a constant state; this mode is used to read out the results after computation or to initialize the cell for a new computation. When the switch is closed, the state integrator receives a current input signal and the cell's state evolves and performs a computation.

The correct initial value and input pixel information are stored during the initialization phase through the I/O MUX multiplexer circuit. Its implementation and operation is discussed in section 4.7.4.

4.5.2 Bias generation

In the mathematical analysis of CNN's the corner points of the output nonlinearity determine the unit for the signals. Thus in the circuit implementation the state-to-output converter properties determine the values of I_{UNIT} and V_{UNIT} (see figure 4.13(b)). The voltage at which the output of the state-to-output saturates determines the unit voltage; this voltage is determined by the gate-overdrive voltage of the differential pair input devices and $V_{UNIT} = \sqrt{2} \cdot (V_{GS} - V_T)_{in}$.

The communication signals between the cells are currents and the state and input signals are voltages so that the connection weights have the physical dimension of conductance or 1/resistance. In the theoretical model the loss term is normalized to 1 and all template weight values are referenced to the

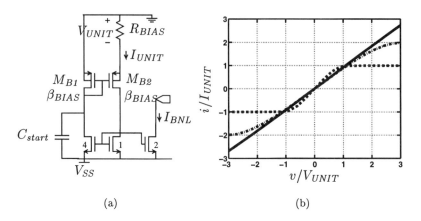

Figure 4.16. (a) Biasing circuit to derive the bias current for the state-to-output converter and the input buffer from the cell resistor. (b) Simulation of the tuning of the characteristics of the cell resistor (-), the state-to-output converter (–) and the input buffer circuit (.-) thanks to the biasing derived by the circuit in (a); all circuit characteristics cross at $[V_{UNIT}, I_{UNIT}]$.

loss term. So for the circuit implementation the value of the cell conductance is the reference or unit value for the weight conductances.

In the ideal model, the slope of the output non-linearity is equal to the loss term; this implies that for the circuit implementation, the following relation must hold:

$$R_{cell} = \frac{V_{UNIT}}{I_{UNIT}} \quad (4.32)$$

The cell resistor characteristic must thus match to the state-to-output converter. Due to process variations, systematic deviations between the properties of these circuits occur so that an on-chip tuning circuit must be designed to derive the bias current for the state-to-output converter from the cell resistance so that (4.32) is satisfied.

We discuss the biasing of the state-to-output converter with current output (figure 4.13(a)) and the input buffer with current output (figure 4.14(a)) in detail; the biasing strategy for the other cell implementations uses the same strategy and similar sizings for the devices. For the state-to-output converter with current output, the maximal output current is the bias current I_{BNL}, which is a DC shifted version of the output and thus $I_{UNIT} = I_{BNL}/2$. Using the definition of V_{UNIT} and I_{UNIT} the following relation is obtained:

$$R_{cell} = \frac{\sqrt{2}(V_{GS} - V_T)_{NL}}{I_{BNL}/2} \tag{4.33}$$

The tuning circuit in figure 4.16(a) derives a bias current so that the g_m's of transistors M_{B1} and M_{B2} are matched to the conductance G_{BIAS} [Kin 94a] [Gre 86]. Transistors M_{B1} and M_{B2} have equal sizes but M_{B1} carries 4 times the current of M_{B2} so that its $(V_{GS} - V_T)$ is twice as large; this implies that the voltage across G_{BIAS} is equal to $(V_{GS} - V_T)_{B2}$ and thus:

$$\frac{1}{G_{BIAS}} = \frac{(V_{GS} - V_T)_{B1}}{I_{UNIT}} \tag{4.34}$$

By using a current factor β_{BIAS} for the M_{B1} and M_{B2} which is half the current factor β_{NL} of M_{NL1} and M_{NL2} in the state-to-output converter, $(V_{GS} - V_T)_{B1}$ is $\sqrt{2}(V_{GS} - V_T)_{NL}$ for the same bias current. When the cell resistor circuit of figure 4.12(a) is used to implement G_{BIAS}, equation (4.34) transforms into equation (4.33) and the biasing circuit ensures a correct tuning of the state-to-output converter to the cell resistor.

The tuning circuit has a second stable operating point for which the devices carry no current; to avoid this unwanted behavior a start-up capacitor C_{start} is provided which forces a large current through M_{B1} when the supply voltage is turned on.

The input buffer circuit must have a linear behavior for the unit input voltage range [-V_{UNIT},V_{UNIT}]; this can be achieved by making the $(V_{GS} - V_T)$ of the input transistors of the input buffer circuit considerably larger than V_{UNIT} and at the same time the transconductance of the voltage to current conversion can be made equal to G_{cell}; this is achieved by making the current factor of the input devices 4 times smaller than of the input devices of the state-to-output converter and by using a 2 times larger bias current.

The bias circuit provides a reference of the unit current I_{UNIT}. Since the resistor in the bias circuit is the cell resistor and carries a current I_{UNIT}, the voltage across the resistor in the bias circuit is a reference for the unit voltage V_{UNIT}. These references are used to derive the correct tuning voltages for the analog weight multipliers in section 4.5.3.

In table 4.4 the sizing and biasing of the different circuits is summarized. In figure 4.16(b) the results of a simulation of the bias circuit, the state-to-output converter, the input buffer and the cell resistor is represented. At the unit voltages V_{UNIT} and $-V_{UNIT}$ the magnitude of the output current of the state-to-output converter and of the input buffer and the current through the cell resistor are all equal to I_{UNIT}. The transconductance of the input buffer is

Table 4.4. Summary of the sizing and biasing of the different CNN cell circuit blocks to obtain a tuning of the circuit block characteristics and properties.

	State-to-output Fig. 4.13(a)	Input buffer Fig. 4.14(a)	Bias circuit Fig. 4.16(a)
Current	I_{BNL}	$I_{BLIN} = 2 \cdot I_{BNL}$	$I_{UNIT} = I_{BNL}/2$
Size	β_{NL}	$\beta_{LIN} = \beta_{NL}/4$	$\beta_{BIAS} = \beta_{NL}/2$
$(V_{GS} - V_T)$	$(V_{GS} - V_T)_{NL}$	$2\sqrt{2}(V_{GS} - V_T)_{NL}$	$M_{B1} : \sqrt{2}(V_{GS} - V_T)_{NL}$ $M_{B2} : 2\sqrt{2}(V_{GS} - V_T)_{NL}$
Transcond.	$g_{mNL} = \sqrt{2} G_{cell}$	$g_{mLIN} = G_{cell}$	$G_{bias} = G_{cell}$

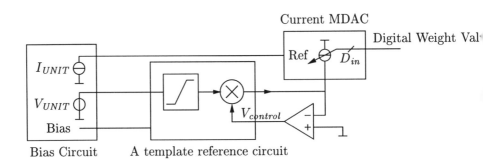

Figure 4.17. Block diagram of the weight tuning system that derives the correct $V_{CONTROL}$ signal from a digital user specified value and from the reference signal provided by the bias circuit in figure 4.16(a).

also constant of the unit range and equal to the conductance of the cell resistor.

4.5.3 Weight tuning strategy

The control voltages $V_{CONTROL}$ of both proposed multipliers (figures 4.5(a) and 4.6) to achieve a wanted multiplication factor depend on the characteristics of the devices and are thus process dependent. Therefore an on-chip weight-tuning system is included to derive the control voltages of the A-template and B-template weight multipliers. Since all cells use the same template settings, a single weight-tuning circuit per template element is sufficient for the whole chip, independent of the number of cells in the array.

If we want to implement an A-template weight of a, this means that for a state voltage $\geq V_{UNIT}$ the output current of the A-template multiplier must be $a \cdot I_{UNIT}$. The output current of the A-template multiplier for a given state voltage depends on the characteristics of the state-to-output converter and the template multiplier circuit. The biasing of the state-to-output converter is already matched to the cell resistance by the current biasing circuit of figure 4.16(a); if a voltage output (figure 4.13(c)) is used however, a variation in the value of the load resistor results in a variation of the output current. Therefore the weight-tuning circuit must include the state-to-output converter and the template multiplier circuit.

In figure 4.17 the block diagram of the weight tuning system is presented for the A-template control voltages. The A-template reference circuit consists of a replica of the state-to-output converter and of a weight multiplier. At the input the unit voltage V_{UNIT} or a voltage larger as V_{UNIT} is applied. The user specifies the wanted template value in digital format (D_{in}). From the reference of I_{UNIT} in the bias circuit of figure 4.16(a) the correct output current is computed with a multiplying current digital-to-analog converter, which uses I_{UNIT} as a reference. This reference output current is forced into the weight multiplier by the opamp, which derives in this way the correct $V_{control}$.

For the B-template multipliers a similar weight-tuning strategy is used. Only the reference circuit is changed into a B-template reference circuit which comprises of a replica the input buffer and of a weight multiplier. At the input a voltage equal to V_{UNIT} is applied.

In this way all process variations in the characteristics of the different building blocks are compensated by the biasing circuit or the weight-tuning system. The circuit designer has however to take into account the possible variation of the $V_{control}$ voltage due to the process variations and has to assure that under worst-case parameter variations, all circuits still function correctly. Although all internal signals of the CNN chip are analog signals, the template or instruction interfacing for the user is completely digital.

4.6 4X4 FULLY PROGRAMMABLE CNN PROTOTYPE CHIP

In this section we present a fully programmable 4x4 CNN chip prototype. It proves the feasibility of the analog VLSI cell architecture presented in section 4.5. This chip is the first published realization of a fully continuous-range programmable CNN in open literature [Kin 94e, Kin 95].

158   ANALOG VLSI INTEGRATION OF MASSIVE PARALLEL SYSTEMS

4.6.1   Cell and chip architecture and implementation

The cell implementation is based on the programmable current mirror circuit of figure 4.5(a) and on the circuits of section 4.5. The full circuit schematic of the transistor realization of a CNN cell is shown in figure 4.18. The state-to-output converter (fig. 4.13(a)) and input buffer (fig. 4.14(a)) have a current output, and these current are mirrored by programmable current mirrors. At the input of the mirror two switches and an inverter are added to select the sign of the connection. To control a connection between two cells, two global control lines are available: one digital control line to select the sign, and one analog control line to program the weight value.

The state capacitor and input capacitor are realized with the gate capacitance of MOS transistors and the cell resistor is implemented as a parallel connection of a nMOS and a pMOS transistor biased in the linear region (fig. 4.12(a)). An on-chip current bias circuit (fig. 4.16(a) is included to derive the correct bias current for the state-to-output and the input-buffer from the cell resistance. Since the programmable current mirror circuits can only realize factors smaller than 1, and since the desired value range is from 1/4 to 4, 4 parallel connected cell resistors are used as the R_{BIAS} in the current bias circuit. Then the bias currents of the state-to-output converter and the input-buffer are multiplied by 4 and their output current at $\pm V_{UNIT}$ is 4 times I_{UNIT} so that the maximal achievable weight value is 4. On-chip an A-template and B-template reference circuit are provided for the weight tuning system. The $V_{CONTROL}$ voltage is routed to a bonding pad and available externally for flexible testing.

In order to monitor the internal evolution of the cells a separate connection to every cell capacitor is provided. Also a direct connection to the input of every cell is included. The I/O circuitry is thus reduced to two switches, and a global control line IO_CONT, but a large number of input (16) and state pins (16) must be provided. For initialization and input the state capacitor is disconnected with the start switch; after initialization, the start switch is closed by the START control signal and the network starts its computation operation.

4.6.2   Circuit sizing and optimization

Choice of V_{UNIT} and I_{UNIT}. The 4x4 CNN prototype chip has been designed for a 2.4 μm standard digital CMOS technology. The supply voltages in this technology are +5/-5. For the sizing of the different circuits, the values for V_{UNIT} and I_{UNIT} have to be selected or optimized.

A typical maximal value for the ratio of the signal range of the state voltage to the V_{UNIT} is 16 (see section 1.5). All circuits connected to the state node, operate correctly for almost the whole supply range, so that the allowed state

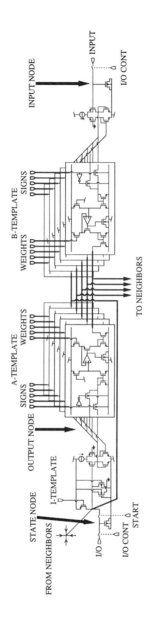

Figure 4.18. Full circuit schematic of a CNN cell of the 4×4 programmable CNN prototype chip.

voltage range is +5/-5; the corresponding value of V_{UNIT} for a state dynamic range of $16 \cdot V_{UNIT}$ is $5\ V/16$ or $V_{UNIT} \approx 300\ mV$.

For the sizing of the programmable current mirror circuit, the choice of the maximal v_{DSmax} over the triode-biased transistors is an important design parameter (see section 4.3.3.A). A large v_{DSmax} causes a large maximal distortion error as given by equation (4.16) and a small v_{DSmax} makes the circuit very sensitive to mismatch errors as shown in equation (4.17). For a correct operation of the CNN, the systematic distortion error as well as the random mismatch error must be small enough. For a v_{DSmax} of 200 mV, the systematic error is 1 % for the largest weight value (4) and 18 % for the smallest weight value (1/4) for the power supply voltages of +/-5 V and the V_T of 800 mV of the 2.4 μm technology. For the transistor sizes used in this implementation, a typical estimated offset voltage for the opamp is about 8 mV, whereas the estimated $(\sigma(\Delta\beta)/\beta)$ and the estimated $\sigma(\Delta V_T)$ for the triode transistors are respectively 0.5 % and 5 mV [†]. This yields an almost constant accuracy of the programmable current mirror of 4 % for a v_{DSmax} choice of 200 mV, which ensures enough yield for a small 4x4 network.

With V_{UNIT} fixed and the value of v_{DSmax} in the programmable current scaling circuit fixed, the only free design parameter is the value of I_{UNIT}. This parameter can be optimized towards a minimal total transistor area in the cell circuit. In order to guarantee a high yield for the network chips, this optimization should take the dependence of the accuracy on the transistor sizing into account. However, for the 2.4 μm technology no technology matching data is available; the minimal used sizes of critical transistors have therefore been made larger than the minimal technology sizes; this intuitive approach is sufficient for a small 4x4 network chip, but for larger chips the accuracy constraint specifications have to be taken into account rigorously (see section 4.7).

The total area of the cell is mainly determined by the area of the A and B template multipliers and their area is determined by the area of the boosting opamps. These opamps (fig. 4.5(b)) are designed for minimal offset voltage; the input $(V_{GS} - V_T)$ is fixed to a small value (200 mV) and the current mirror $(V_{GS} - V_T)$'s are fixed to a large value (500 mV) (see chapter 2), such that the input range includes the V_{DD}. In this way the $(V_{GS} - V_T)$ of all transistors is fixed.

The size of a transistor for a given fixed $(V_{GS} - V_T)$ depends on the bias drain-source current I_{DS} through the transistor. From (A.20) we calculate

---

[†]these numbers have been estimated using mismatch technology constants of a similar CMOS technology in table 2.1.

that the $\left(\frac{W}{L}\right)$ of a transistor is in first order given by:

$$\left(\frac{W}{L}\right) = \frac{2I_{DS}}{\mu C_{ox}(V_{GS} - V_T)^2} \qquad (4.35)$$

For small I_{DS} values ($< \mu C_{ox}(V_{GS} - V_T)^2/2$) a minimal transistor width W_{min} is chosen to minimize the transistor gate area and the transistor length is $W_{min}/\left(\frac{W}{L}\right)$; the area of the transistor is then $W_{min}^2(\mu C_{ox}(V_{GS} - V_T)^2)/2I_{DS}$ and is inversely proportional to the bias current. For large I_{DS} bias values ($> \mu C_{ox}(V_{GS} - V_T)^2/2$) a minimal length L_{min} transistor is used and the area is proportional to the bias current: $L_{min}^2 2I_{DS}/(\mu C_{ox}(V_{GS} - V_T)^2)$. The area of the transistor as a function of the bias current thus has a U-shaped characteristic – like the characteristic in figure 4.19 – and the optimal current for a minimal area is $\mu C_{ox}(V_{GS} - V_T)^2/2$; then the transistor has a minimal width and length.

The OTA bias current I_{BIAS} (figure 4.5(b)) is derived from I_{UNIT} current reference, so that the bias currents through the transistors are proportional to I_{UNIT}. From the discussion in the previous paragraph we know that if the I_{UNIT} is increased, the transistors in the opamp must be made wider to keep their $(V_{GS} - V_T)$ optimal and the area of the opamp increases proportional to I_{UNIT}. On the other hand if the I_{UNIT} is strongly decreased the transistors become longer to keep their $(V_{GS} - V_T)$ correct. The dependence of the total area of the opamp is a summation of the U-shaped characteristics of the individual transistors. This summation yields again a U-shaped dependence of the the total area of the opamp on I_{UNIT} so that an optimal value in I_{UNIT} exists for which all transistors are about minimal size and so that the area of the opamp is minimal.

Since the v_{DSmax} in the programmable mirror and the V_{UNIT} are fixed, the $(V_{GS} - V_T)$ of all transistors in the A and B template circuits is fixed. Also for the transistors in the state-to-output converter, the input buffer and the cell resistor, the $(V_{GS} - V_T)$ values are fixed by the choice of V_{UNIT} and the biasing relations of table 4.4. Consequently the dependence of the total cell circuit is a summation of the U-shaped dependencies of the individual transistors and also has a U-shape (see figure 4.19). The only remaining free parameter is the ratio between the I_{UNIT} and the bias current I_B of the opamp; this factor has been optimized using a spreadsheet sizing model for the cell circuits. The opamp is best biased with a current 5 times smaller as the state-to-output converter or $I_{BIAS} = 8/5 I_{UNIT}$. In figure 4.19 the U-shaped dependence of the total cell transistor area on the choice of I_{UNIT} is plotted. The optimal value for I_{UNIT} is 1.25 μA in this technology. All design parameters are now fixed. Their

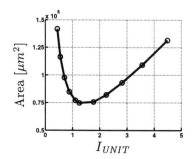

Figure 4.19. The value of I_{UNIT} is optimized to obtain a minimal total active area of the transistors in the cell. The wiring area is not included.

relation to the W/L ratios of all transistors can be derived from the equations in the previous sections.

In table 4.5 the signal-transistor sizings and biasing of the different cell building blocks is summarized. The total cell circuit contains 218 transistors.

Circuit simulation results. The simulated operation of the state-to-output converter followed by a programmable current mirror is illustrated in figure 4.20(a). A negative sign is selected for the connection, and the transfer characteristic is controlled by the $V_{CONTROL}$ voltage. The V_{UNIT}, the state voltage at which the saturation in the output occurs is independent of the template weight selected and is at 300 mV.

The $V_{CONTROL}$ voltage is in first order linearly dependent on the desired template weight; in figure 4.20(b) the template value is plotted as a function of the control voltage $V_{CONTROL}$ as it is derived by the weight tuning system. For smaller factors the influence of the distortion errors increases; the weight tuning system compensates them partially and the weight tuning characteristic becomes slightly non-linear.

The synapse circuits contain a feed-back loop containing the boosting OTA. The OTA is of the load-compensated type and the stability of the loop has been assured by an extra compensation capacitor ($C_{comp} = 0.2\ pF$) at the output of the OTA (see figure 4.5(b)) in the synapses to achieve a phase margin of 60 degrees for a GBW of 4 MHz for the loop including the cascode transistor M_4 (see figure 4.5(a)).

The smallest cell time constant for which the cell will still operate correctly is, must be at least 10 times larger as the parasitic time constant in the synapse circuits; the parasitic time constant in the synapses is determined by the step response of the feed-back loop and thus the GBW of the OTA; the worst case parasitic time constant is lower than 1 μs so that the cell time constant has

Table 4.5. Summary of the biasing and the sizing of the building blocks of the 4x4 CNN prototype chip for the 2.4 μm CMOS technology.

V_{UNIT}	300 mV	TECHNOLOGY				
I_{UNIT}	1.25 μA	K_{Pn}	50 $\mu A/V^2$			
C_{state}	40 pF	K_{Pp}	20 $\mu A/V^2$			
		CELL RESISTOR				
R_{cell}	240 $k\Omega$	$\left(\frac{W}{L}\right)_N$	$\left(\frac{2.4}{156}\right)$	$\left(\frac{W}{L}\right)_P$	$\left(\frac{2.4}{42.5}\right)$	
		STATE-TO-OUTPUT CONVERTER				
I_{BNL}	$8 \times 1.25 \mu A$	$\left(\frac{W}{L}\right)_{1,2}$	$4 \times \left(\frac{14}{5.2}\right)$	$\left(\frac{W}{L}\right)_{3,4}$	$\left(\frac{7.2}{5.2}\right)$	
		INPUT BUFFER				
I_{BLIN}	$16 \times 1.25 \mu A$	$\left(\frac{W}{L}\right)_{1,2}$	$\left(\frac{14}{5.2}\right)$	$\left(\frac{W}{L}\right)_{3,4}$	$\left(\frac{14.4}{5.2}\right)$	
		A TEMPLATE MULTIPLIER				
		$\left(\frac{W}{L}\right)_{1,2}$	$\left(\frac{4.8}{20.8}\right)$	$\left(\frac{W}{L}\right)_4$	$\left(\frac{7.2}{2.4}\right)$	
		B TEMPLATE MULTIPLIER				
		$\left(\frac{W}{L}\right)_{1,2}$	$\left(\frac{4.8}{11.2}\right)$	$\left(\frac{W}{L}\right)_4$	$\left(\frac{13.6}{2.4}\right)$	
		TEMPLATE BOOSTING OTA				
I_{BIAS}	$8/5 \times 1.25$ μA	$\left(\frac{W}{L}\right)_{1,2}$	$\left(\frac{4.8}{4.8}\right)$	$\left(\frac{W}{L}\right)_{3,4}$	$\left(\frac{4.8}{6.4}\right)$	
$\left(\frac{W}{L}\right)_{5,6}$	$\left(\frac{10.4}{4.8}\right)$	$\left(\frac{W}{L}\right)_{7,8}$	$\left(\frac{2.4}{14.4}\right)$	C_{comp}	0.2 pF	

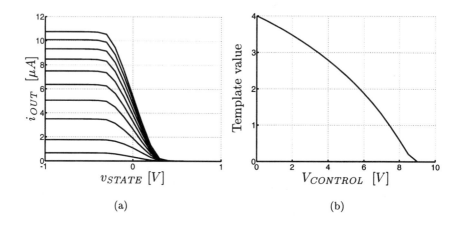

Figure 4.20. (a) Simulation of the operation of the state-to-output converter and an A-template multiplier for a negative connection; (b) the A-template weight factor as a function of the $V_{CONTROL}$ voltage.

been chosen to be 10 μs. A state capacitor of 40 pF has been added at the state node to set the cell time constant; this capacitor is fitted under the global wiring to the cells, so it requires no significant extra layout area.

4.6.3 Chip and Measurement results

Lay-out. The 4x4 CNN chips have been fabricated in the 2.4 μm standard CMOS technology of Mietec-Alcatel; this technology has two metal interconnect levels and one poly layer available and no analog extensions. The connections between the cells have been limited to the north, south, east and west neighbors, which is sufficient for a very large subclass of templates. Figure 4.21 is a micro-photograph of the chip; the 4 rows and 4 columns of cells can be clearly distinguished. At the bottom a row with 4 circuits to set the edges for the network have been included. In the upper left corner an extra cell has been laid out; it serves as part of the weight tuning system and is used to test a cell circuit individually. The total chip contains over 4600 active devices. Since a direct connection to the state node and the input node of every cell is provided a large number of bonding pads are required. The chips have been packaged in a standard 84 pins PLCC package.

Measurement set-up. A dedicated measurement set-up has been developed to test the chips and to demonstrate their operation. The operation of the measurement system is illustrated in figure 4.22. The user composes the input

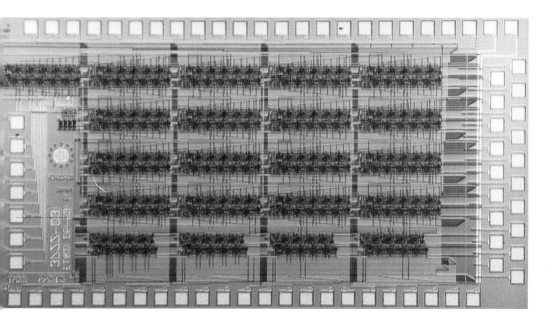

Figure 4.21. Micro-photograph of the analog 4x4 fully programmable CNN prototype chip in a standard 2.4 μm CMOS technology.

166   ANALOG VLSI INTEGRATION OF MASSIVE PARALLEL SYSTEMS

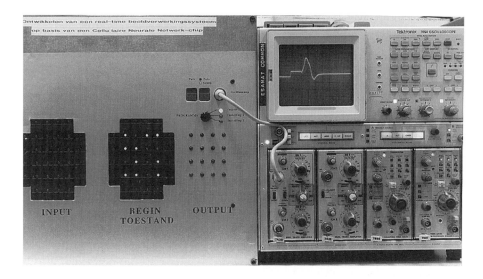

Figure 4.22. Photograph of the test- and demonstration set-up for the 4x4 CNN prototype chip; the input and initial state image are composed on the front panel and the resulting output image is displayed on the outer-right LED display; the internal state evolutions of the cells can be monitored with an oscilloscope.

image and the initial state image on the front panel; then the computation is executed by the chip and the result is displayed on the leftmost diode screen. Three sets of template settings can be preprogrammed and the user can select between them. The set-up contains circuitry to transfer the images to the inputs and the states of the cells, to generate the control signals to execute the template, and circuitry to read out the results and display them on the output screen; extra circuitry is provided to follow and buffer the internal state evolution of a cell and display it on an oscilloscope.

The 4x4 CNN chips have been tested with a large number of templates. Using the connected component detector template (CoCoD) settings of table 1.2, the network counts the number of horizontal connected pieces in the image. Each piece is reduced to a size of one pixel and these pixels are shifted to the right. In figure 4.23(a) an input image and its measured output image is displayed. The input image is used as the initial state for the network. The transient evolution of the states of the cells of the top row is given in fig-

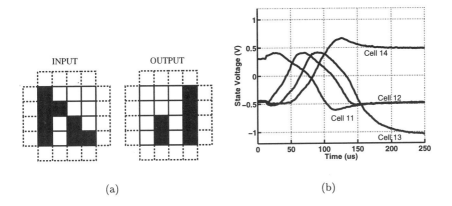

Figure 4.23. (a) input and measured output image for a horizontal connected component detector to the right; the dotted lines represent the edges which remain constant. (b) Measured transient evolution of the state voltages of the top row of cells in (a).

ure 4.23(b). A voltage above $+V_{UNIT}$ ($+0.3\ V$) results in an output of $+1$ or a black pixel and a voltage below $-V_{UNIT}$ ($-0.3\ V$) is a white pixel. After the initialization of about 20 µs the self-feedback in the cells is closed with the START switch and the network starts its evolution towards equilibrium. The top row contains only one connected component at the out-most left position and the shifting of this pixel to the right can be observed. A similar functioning of the chip can easily be obtained for a horizontal detector to the left or for a vertical detector by rearranging the template values. The connected component detector function demonstrates the power of the dynamic nature of the network. The integrated state capacitance C_{state} and cell resistance R_{cell} set the cell time constant ($\tau = R_{cell} \cdot C_{state}$) to 9.6 µs (see table 4.5). From figure 4.23(b) the measured cell time constant is calculated using the fact that the tops of the evolutions of two subsequent cells in a CoCoD are $1.43 \cdot \tau$ apart and a value of 17.5 µs has been measured. This difference is due to the extra capacitive loading of the state nodes by the measurement setup and an appropriate buffering and I/O scheme can solve this problem.

When the chip is configured with the holefiller template (HOLE_MOD) of table 1.2, it performs a hole-filling operation. For this template the input image is put on the input capacitors of the cells and the states are all initialized to black ($+V_{UNIT}$). Figure 4.24(a) represents an input image, containing a hole, and its measured output image with the hole filled. The transient evolution of representative cells in the image are given in figure 4.24(c). The other cells have a similar behavior due to the symmetry in the image. In the input image

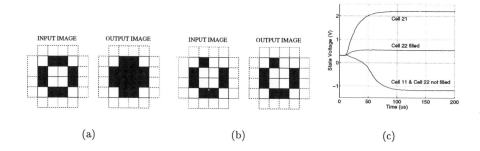

Figure 4.24. (a) Input image for the holefiller containing a hole and the corresponding measured output image with the hole filled. (b) Input image without holes; the measured output image is the same as the input image. (c) Measured transient evolution of a few typical cells in the holefiller.

of figure 4.24(b) the hole is opened and it is not filled by the network. The transient evolution of the cells that are cleared is also displayed on figure 4.24(c). By using the edges of the network the input images of figure 4.25(a) and 4.25(b) are constructed. The propagation of information through the image can be observed: all cells start from black; white propagates from the right edge until it hits a black cell, which is visible in the evolution of the states in figure 4.25(c). In figure 4.25(a) the spiral is fully cleared while in figure 4.25(b) the white stops at cell 41 and the hole in the image remains filled. The hole-filler template demonstrates that global operations can be executed on the image by exploiting the propagation of information through the network.

Performance. The programmable current scaling circuit including the OTA and the sign switches occupies $52\lambda \times 47\lambda$ or 14000 μm^2 in the 2.4 μm technology, and the OTA consumes half of this area. The total cell size including routing is 410 $\mu m \times 800$ μm, which corresponds to a cell density of 3 $cells/mm^2$.

The speed of computation is determined by the time constant of the cells ($\tau = R_{cell} \cdot C_{state}$), which is 9.6 μs (see table 4.5). The power consumption of the cells depends on the used template. The power consumption of the state-to-output converter, of the input buffer, and of the boosting opamp and the input stage of the programmable current mirror circuit is fixed; the DC current through the output stage of the programmable current mirror is dependent on the weight and is $|A_i| \cdot I_{UNIT}$ for an A-template multiplier implementing a weight A_i and $|B_i| \cdot 2I_{UNIT}$ for an B-template multiplier implementing a

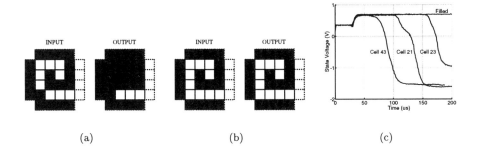

Figure 4.25. (a) Spiral input image, using the edge cells, for the holefiller without holes and the corresponding measured output image. (b) Input image with partial spiral and hole. In the measured output image this hole is filled. (c) The transient traces of the cells show the propagation of white from the right edge.

weight B_i. The total DC current flowing into a cell is then calculated as:

$$I_{DC} = \left(\sum_i |A_i| + \sum_i 2 \cdot |B_i| \right) \cdot I_{UNIT} \qquad (4.36)$$

This DC current is added to the current through the I-template bias transistor and it is also the maximal amplitude of the negative I-template value that can be generated, since only a pMOS transistor is used for the I-template bias. For all templates from [Ros 94] this poses no problem. The current consumption of the active circuits of a connected component detector cell is $40 \cdot I_{UNIT}$ or $50 \ \mu A$, and the holefiller cell consumes $107 \cdot I_{UNIT}$ or $134 \ \mu A$ from a -5/+5 power supply. Since all building blocks are current biased the power consumption of the chip is well controlled; it only depends on the value of I_{UNIT} which is generated on-chip and will slightly vary due to process variations. The experimental obtained values for the unit current I_{UNIT} and the power consumption from the test-chips agree well with the calculations.

4.6.4 Evaluation

This prototype chip has demonstrated the feasibility of the realization of fully programmable CNN chips using analog circuitry in a standard digital CMOS technology. The performance of the chip is summarized in table 4.6. The cell architecture, the bias circuitry and the weight tuning system all have demonstrated satisfactory operation. The main shortcomings of this design are its simple I/O structure, which is only feasible for a very small number of cells. Also the lack of detailed matching information of the transistors in the used

170   ANALOG VLSI INTEGRATION OF MASSIVE PARALLEL SYSTEMS

Table 4.6. Summary of the experimental performances of the 4x4 CNN prototype chip.

CELL DENSITY		
2.4 μm CMOS	3 cells/mm^2	
SPEED		
Cell time constant	10 μs	
POWER CONSUMPTION		
Connected Component detector	50 μA/cell	(500 μW)
Holefiller	134 μA/cell	(1340 μW)
PROGRAMMABILITY		
A & B template weight values	$\pm 1/4$ to ± 4	

technology, restrict the circuit sizing optimization and forces to use a reasonable approximation and extrapolation of the matching quality numbers for the technology.

4.7 20X20 ANALOG PARALLEL ARRAY PROCESSOR

In this section, we present a full sensor signal processing system based on a 20x20 analog parallel array processor (APAP). The computing core of the chip is a fully programmable continuous time CNN array of 400 cells; the template weights are programmable over a continuous value range from 1/4 to 4 and the signs of the connections are user-selectable; the cells operate in continuous time and in parallel.

An I/O interface system is provided that allows the direct coupling of 2D array sensors to the chip. The chip is further interfaced to a personal computer (PC). The signals from the sensors are fed directly into the APAP and the results of the computation are sent to the PC.

The main contributions of the VLSI design of this system are:

- the use of high density linear transconductance template multiplying circuits;

- the design of an efficient I/O system that allows the direct coupling of 2D array sensors to the chip and an interface to a PC;

- the rigorous design method taking into account the sensitivity of the correct system operation to device inaccuracies through design specifications;

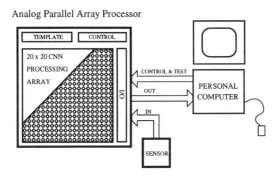

Figure 4.26. Block diagram of the sensor signal processing system built with the APAP chip.

- the circuit sizing optimization using quantitative technology matching data towards a maximal circuit density while guaranteeing a high chip yield.

Thanks to its efficient interface for array sensors, its high speed parallel analog computations and its full programmability, this chip can be called an analog parallel array processor (APAP).

4.7.1 System architecture

Figure 4.26 represents the block diagram of he realized parallel signal processing system for real-time sensor interfacing. The analog parallel array processor (APAP) chip is the core of the system. The sensor provides the analog input signals for the APAP chip. An interface-card links the APAP chip to a personal computer (PC) that is used for control, memory, user interfacing, and testing.

4.7.2 Cell architecture and circuit implementation

The full cell circuit implementation is represented in figure 4.27. The cell block diagram is based on the linear grounded transconductance multiplier of figure 4.6. The A-template multipliers are driven with a nMOS input transistors version of the state-to-output converter with voltage output in fig. 4.13(c); it converts the single ended state voltage non-linearly in the differential output voltage. The B-template multipliers are driven by a nMOS input transistors version of the input buffer with extended linear range and voltage output in fig. 4.14(b); it converts the single-end input voltage into a differential voltage. The input storage capacitor and the cell state capacitor are implemented as transistor gate capacitors; the cell resistor is implemented with a parallel connection of an nMOS and a pMOS transistor biased in the linear region (see fig. 4.12(a)). Furthermore a current buffer (see fig. 4.15) is included between the current input of the cell and the state node; it limits the swing on the current summing input so that no output impedance buffering is necessary in the individual template multipliers.

4.7.3 Circuit sizing and optimization

The circuit has been designed for a 0.7 μm standard digital CMOS technology from Mietec-Alcatel. The chip is designed for a -2.5/+2.5V power supply. The technology has analog extensions but these have not been used for this design.

The circuit sizing has been done for a specified speed of the cells or cell time-constant and the chip is designed for a high yield. Consequently the accuracy of the template circuits and the location of the second pole in the cell circuit are fixed specifications in the design. The free parameters in the design are then used to minimize the cell area and thus maximize the cell density. In the forthcoming paragraphs the sizing optimization of the cell building blocks is discussed in detail. The circuit sizing of the building blocks always follows the same computational path. First the (W/L)'s or β's ($\beta = \mu C_{ox}(W/L)$) of the transistors are calculated from the conditions for correct biasing and correct circuit operation. The areas of the devices $(W \cdot L)$ remain free parameters that directly determine the matching of the device parameters and thus the accuracy of the template values; the device areas are minimized with a constrained optimization method that takes into account the required accuracy specifications to obtain a high chip yield. The speed requirements are achieved by optimizing the driving capability of the state-to-output converter to position the second pole time constant (see below). Once the (W/L)'s and the areas are known the transistor sizes W and L are easily calculated.

In the following paragraphs we discuss this procedure for each of the building blocks in detail but first we discuss the choice of V_{UNIT} and I_{UNIT}.

Choice of V_{UNIT} and I_{UNIT}. The power supply voltage of this technology is 5 V. For a programmable CNN chip implementation the largest possible range in the state voltage must be designed for so that the unit voltage has been chosen as small as possible. To obtain a good saturating output transfer characteristic in the state-to-output converter, its input transistors have to be biased in strong inversion i.e. with a $(V_{GS} - V_T) \geq 150\ mV$; the minimal possible unit voltage V_{UNIT} is $\sqrt{2}$ larger and is fixed to be 210 mV.

The design of all circuits is determined by the choice of I_{UNIT}; the output current and thus the size of the input devices of the template multipliers is proportional to I_{UNIT} and the biasing of the state-to-output converter and input buffer is related to I_{UNIT} to obtain a good matching of the template values to

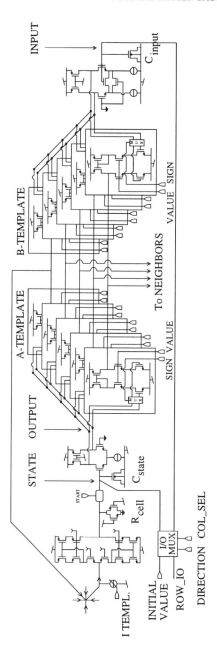

Figure 4.27. Full circuit schematic of a CNN cell of the 20x20 APAP chip.

the cell resistance under process variations as explained in section 4.5.2. The choice of the I_{UNIT} thus strongly influences the overall design and its value is basically a free design parameter that is optimized to minimize the cell area (see further).

Sizing of the template multipliers. The output current of the template multipliers is given by equation (4.19) so that the template value is $\beta_{1,2} V_{inMAX} V_C / I_{UNIT}$.
The specified programming range for the template values is from 1/4 to 4. The input transistors of the multipliers are then sized by solving the following equation for $\beta_{1,2}$:

$$A_{MAX} = 4 = \frac{\beta_{1,2} v_{inMAX} V_{CMAX}}{I_{UNIT}} \tag{4.37}$$

The (W/L) of the input transistors is thus fixed if v_{inMAX}, V_{CMAX} and I_{UNIT} are known.

The correct biasing and the accuracy of the template values provide the relations for the determination of V_{CMAX} and v_{inMAX}. For correct biasing V_{CMAX} and v_{inMAX} must satisfy the constraint in (4.20):

$$v_{inMAX} + V_{Cmax} + V_{T3,4} \leq (V_{DD} - V_{SS}) \tag{4.38}$$

The relative accuracy of the output current of the template multiplier, which influences the accuracy of the template value (see below), is expressed by (4.26):

$$\left(\frac{\sigma(I)}{I}\right)^2_{out} = \left(\frac{\sigma(V_{T1,2})}{v_{in}}\right)^2 + \left(\frac{\sigma(\beta)}{\beta}\right)^2_{1,2} \left(\frac{V_{CM} - V_{T1,2} - V_C/2}{v_{in}}\right)^2$$
$$+ \sigma^2(M) \left(\frac{1}{2} + \frac{V_{CM} - V_{T1,2} - V_C/2}{v_{in}}\right)^2 \tag{4.39}$$
$$+ \sigma^2(V_{C1,2}) \left(\frac{1}{v_{in}} + \frac{V_{CM} - V_{T1,2} - V_C/2}{v_{in} V_C}\right)^2$$

We see that the accuracy is also dependent on the choices of the signal ranges: the larger the V_{inMAX} the lower the sensititivy to input transistor mismatches, but the lower the allowed V_{CMAX} and thus the higher the sensitivity to cascode transistor mismatches. The values of V_{inMAX} and V_{CMAX} are thus free parameters for the design of the multipliers but they are constrained by (4.38) and (4.39).

The $(V_{GS} - V_T)$ of the current mirror transistors M_5-M_6 is fixed to 0.5 V to obtain accurate current mirroring (see chapter 2) and their (W/L) is computed from the maximal current flowing through the input devices which is calculated using (4.18) for a maximal weight and input signal. The (W/L) of the cascode transistors M_3-M_4 is specified as at least 10 times larger than the (W/L) of the input transistors to limit the distortion error.

From these relations the (W/L)'s of all transistors can be calculated. The area of the devices influences the matching of the devices and this influences the accuracy of the output current in (4.39); the area of the devices are free design parameters, that are optimized to satisfy the accuracy requirements.

Cell time constant and bias of state-to-output converter. In order to optimize the speed of the computations, the cell time constant must be made as small as possible. The second pole time constant in the cell implementation must however remain 10 times smaller than the cell time constant, in order to avoid erroneous cell evolutions (see chapter 3).

The second pole in the cell circuit is located at the output node of the state-to-output converter (see figure 4.27); the output resistor of the state-to-output converter, which is approximately equal to R_{LOAD}, together with the input capacitance of the five A-template synapse multipliers determine the second pole time constant. The speed performance of the cell is a design specification and the wanted cell time constant is 2.7 μs so that the parasitic second pole time constant must be smaller than 0.2 μs.

The swing at the output node of the state-to-output converter is determined by the (large) wanted input swing for the template multipliers V_{inMAX} (see above) to achieve a good template accuracy. Therefore a large R_{LOAD} is preferred to reduce the bias current I_{BNL} and thus the power consumption of the state-to-output converter. However, a large R_{LOAD} leads to a slow second pole time constant. The only way to improve the speed of the system is then to lower the R_{LOAD} of the state-to-output converter; the accuracy of the multiplier is fixed by the yield specifications so the load capacitance at the output node is basically fixed (see chapter 2). When the R_{LOAD} is lowered, the bias current must be increased to keep the voltage swing constant; the voltage swing may not decrease because that results in higher mismatch sensitivity (see equation (4.39)) and consequently requires to make the multiplier input devices even larger and thus the parasitic time constant larger again. As a result we can only achieve a set speed specification by choosing a larger bias current for the state-to-output converter.

The bias current I_{BNL} of the state-to-output converter must be derived from the unit current I_{UNIT} for correct weight tuning (see table 4.4); however its value can be further increased to a larger integer multiple of the I_{UNIT} if the input devices width W is scaled with the same factor so that the $(V_{GS} - V_T)$ of the input devices remains constant. This scaling factor or the integer ratio

I_{BNL}/I_{UNIT} is the free design parameter that is optimized to satisfy the speed constraint.

State-to-output converter. In the previous paragraph the determination of the bias current for the state-to-output converted is outlined. Once the bias current is known all transistor (W/L)'s can be determined using equation (4.35), since their $(V_{GS} - V_T)$'s are also known: the $(V_{GS} - V_T)$ of the input transistors M_1-M_2 is determined from the fixed unit voltage V_{UNIT} (see discussion above) and the current source transistor is sized with a $(V_{GS} - V_T)$ of 0.5 V to obtain good current matching. The equivalent resistance of the resistive load transistors M_3-M_4 must be equal to $R_{LOAD} = 2 \cdot V_{inMAX}/I_{BNL}$ and their $(W/L)_{3,4}$ or $\beta_{3,4}$ is determined using the relation $1/R_{LOAD} = \beta_{3,4} \cdot (V_{DD} - V_{SS} - |V_{Tp}|)$ taking into account the mobility reduction due to the high $(V_{GS} - V_T)$ bias [Lak 94].

The mismatches in the transistors results in a variation of the output signal – and thus of the input signal of the template multipliers – which results in a variation of the template output current. This error contributes to the total error of the template value (see below). The relative error in the output voltage which is given by:

$$\frac{\sigma^2(\Delta v_{OUT})}{v_{OUT}^2} = \frac{\sigma^2(\Delta R_{LOAD})}{R_{LOAD}^2} + \frac{\sigma^2(\Delta I_{BNL})}{I_{BNL}^2} \quad (4.40)$$

$$= \frac{\sigma^2(\Delta \beta_{3,4})}{\beta_{3,4}^2} + \frac{\sigma^2(\Delta V_T)}{(V_{GS} - V_T)_{3,4}^2} + \frac{\sigma^2(\Delta I_{BNL})}{I_{BNL}^2} \quad (4.41)$$

and the relative error on the bias current I_{BNL} is computed using equation (2.29) and (2.31). Through the constraint on the template value variation the areas of the devices are optimized (see below).

Linearization input buffer. The cell input signal remains constant during the cell evolution so the speed of the input buffer is not crucial, even though it is also loaded with a high load capacitance of the inputs of the 5 B-template multipliers.

To obtain a tuning or value matching of the cell resistance, the A-template weights and the B-template weights, the bias circuit of figure 4.16(a) has been applied in this chip. It derives the correct biasing current to match the saturation point of the state-to-output converter to the cell resistance value. The input buffer includes a Krummenacher degeneration, which yields a higher linear region for the buffer with only a very slight increase in the bias current. The different device structure of the linear input buffer requires a different biasing than in table 4.4 for the input buffer. A bias current of only $I_{BLIN} = 1.25 \times I_{UNIT}$

> Given: Mismatch models technology
> Circuit & accuracy expressions
> Constraints: Accuracy specifications (Yield)
> Speed specification
> Correct Biasing and Circuit Operation
>
> ⇒ Optimize the signal swings
> the bias points
> the device sizes
>
> ⇒ to Minimize cell area

Figure 4.28. Outline of the constrained optimization problem for sizing the cell circuits; the cell density is maximized while guantee-ing a correct cell operation by controlling the cell accuracy in every step through the accuracy constraints.

is used for each of the transistors. The signal transistors of the state-to-output converter are made 2.5 times wider as the input transistors of the input buffer and the 'Krummenacher' degeneration transistors are 4 times longer. This yields a sufficiently linear buffer with a transconductance equal to $1/R_{cell}$, with only a power consumption increase of 25 %. A detailed sizing plan of this structure can be found in [Kru 88]. It is clear that this linearization method is more power efficient than the one used in the 4x4 prototype chip. The transistors (W/L)'s are now fixed and the areas are minimized using a similar constraint as in (4.40) for the B-template weight accuracies.

Constrained optimization. In the previous paragraph the different biasing and speed constraints are discussed for the transistor (W/L)'s and signal swings in the different building blocks. Also the effect of the transistor areas on the accuracy of the template values is explained. As pointed out in this discussion, several parameters can still be chosen freely by the designer for the sizing of the transistors. This design freedom has been used to minize the total cell area to obtain a maximal cell density. The free design parameters are: the voltage swing on the output node, which is also the maximal input signal for the template multiplier V_{inMAX}; the range of the weight tuning voltage in the template multipliers V_{CMAX}; the bias current I_{BNL} of the state-to-output converter; the unit current I_{UNIT}; the area of the devices.

The total accuracy of the A template values A is determined by the accuracy of the output voltage of the state-to-output converter (4.40) and the template multipliers output current (4.39). From (4.19) it is clear that the contribution in the relative error in the template value due to the mismatches in the state-to-output converter, is equal to the relative error in the converter output voltage; the contribution in the relative error in the template value due to the mismatches in the multipliers, is equal to the relative error in the multiplier output current. All error sources are independent and normally distributed since they are caused by different devices so that the relative variance $\sigma^2(\Delta A)/A^2$ of the total template error is the sum of the variances in (4.39) and (4.40). For the B-template values a similar expression depending on the matching in the linear input buffer and the B-template multipliers can now be easily derived.

In chapter 3 the necessary accuracy specifications to guarantee a high chip yield and a correct system operation have been derived for a 20x20 network and for a large set of template settings. These specifications have been repeated in figure 4.29 and express the required total accuracy of the template multiplier circuit. The area of the devices is optimized using this constraint and the expression for template accuracy derived above. The quantitative matching models and experimental data that have been discussed in chapter 2 for the 0.7 μm technology are used during the sizing to calculate the transistor parameter mismatch dependence on the area.

These accuracy specifications, the speed specification and the correct biasing conditions form the total set of constraints in the constrained sizing optimization problem for this chip design, outlined in figure 4.28. This constrained sizing optimization problem of the total cell circuit is not straightforward and cannot be solved analytically. It has been implemented in a spreadsheet sizing model in EXCEL [Exc] and the free parameters have been optimized towards a minimal cell area using the constrained optimization tool of EXCEL.

Sizing. The area minimization has resulted in a unit current $I_{UNIT} = 2\ \mu A$ and a biasing for the state-to-output converter of $16 \times I_{UNIT}$. In table 4.7 the representative optimized transistor sizes and biasing parameters of the different building blocks are summarized. In figure 4.29 the achieved accuracy for the A and B template circuits is plotted together with the required accuracy specifications. The circuit implementation meets the specifications so that a high circuit yield will be achieved (see chapter 3).

4.7.4 I/O system design

The Input-Output interface system consists of three main parts. A first part is included in every cell; a second part is a digital control block to control the I/O

Table 4.7. Summary of the biasing and the sizing of the building blocks of the 20x20 APAP chip for the 0.7 μm CMOS technology.

V_{UNIT}	210 mV	TECHNOLOGY			
I_{UNIT}	2 μA	K_{Pn}	70 $\mu A/V^2$		
C_{state}	27 pF	K_{Pp}	20 $\mu A/V^2$		
		CELL RESISTOR			
R_{cell}	100 $k\Omega$	$\left(\frac{W}{L}\right)_N$	$\left(\frac{2}{33}\right)$	$\left(\frac{W}{L}\right)_P$	$\left(\frac{2}{9.5}\right)$
		STATE-TO-OUTPUT CONVERTER			
I_{BNL}	32 μA	$\left(\frac{W}{L}\right)_{1,2}$	$\left(\frac{30.6}{1.5}\right)$	$\left(\frac{W}{L}\right)_{3,4}$	$\left(\frac{3}{9.5}\right)$
		INPUT BUFFER			
I_{BLIN}	$1.25 \times 2\mu A$	$\left(\frac{W}{L}\right)_{1,2}$	$4 \times \left(\frac{2.8}{4.5}\right)$	$\left(\frac{W}{L}\right)_{3,4}$	$\left(\frac{2}{21.3}\right)$
C_{input}	9 pF	$\left(\frac{W}{L}\right)_{1b,2b}$	$\left(\frac{2.8}{4.5}\right)$		
		A TEMPLATE MULTIPLIER			
$\left(\frac{W}{L}\right)_{1,2}$	$\left(\frac{2.1}{51.3}\right)$	$\left(\frac{W}{L}\right)_{3,4}$	$\left(\frac{6}{1.5}\right)$	$\left(\frac{W}{L}\right)_{5,6}$	$\left(\frac{4.3}{9.2}\right)$
V_{inPP}	1.8 V	V_{CM}	4.1 V	V_{Cmax}	1.7 V
		B Template Multiplier			
$\left(\frac{W}{L}\right)_{1,2}$	$\left(\frac{2.1}{51.3}\right)$	$\left(\frac{W}{L}\right)_{3,4}$	$\left(\frac{6}{1.5}\right)$	$\left(\frac{W}{L}\right)_{5,6}$	$\left(\frac{4.3}{9.2}\right)$
V_{inPP}	2.25 V	V_{CM}	3.875 V	V_{Cmax}	1.7 V

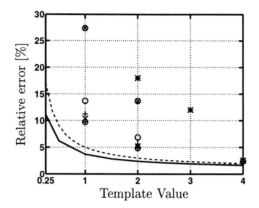

Figure 4.29. The allowed relative error for the template values for a 20x20 network and a yield of 95%: o=CoCoD, x=HOLE_MOD, ⊗=SHADOW, ⊕=EDGE, +=NOISE, and *=PEEL templates. The specification for the programmable template circuit is (--) and the accuracy of the realized circuit is (-).

functions and the third part is located at the bidirectional I/O pins. The I/O system in this chip is based on the X-Y addressing scheme that is discussed in section 4.4.3.2.

Cell Interface. The part that is included in every cell shown on the schematic of figure 4.27. All cells in a row are connected to the ROW_IO line; through this line the input pixels are routed to the cells column by column. The global I/O line INITIAL_VALUE contains the initial state to initialize the cell state capacitors before computation. The I/O MUX connects these two lines to the correct internal cell nodes: the input node or the state node depending on the signal on the control lines COL_SEL and DIRECTION. The COL_SEL signal determines which column in the array of cells is loaded with the sensor signal pixels on the ROW_IO line. The DIRECTION signal determines if the pixels are loaded onto the input capacitor C_{input} or the state capacitor C_{state}, depending on the requirements of the template. The I/O MUX contains the necessary transmission gates and control signal inverters to implement these functions.

Digital Control Block. The speed transfer of signals from the sensor into the array is mainly determined by the speed of the used sensor; therefore the clock for setting the transfer speed is provided from off-chip to the on-chip I/O control block. In order to mark a new input image, a set pulse is provided

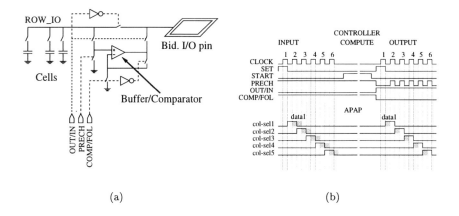

Figure 4.30. (a) Schematic of the bidirectional row I/O pin and circuitry; in the output mode a digital or an analog signal can be buffered and outputed. (b) Control signals generated by the global system controller for the INPUT, COMPUTE and OUTPUT phases in a 5x5 network; on the bottom the internal COL_SEL signals for the different columns are shown and the valid timing for the input and output data is indicated.

on the second control line of the I/O control block; this pulse is issued by the global system controller which is in the present set-up of figure 4.26 the personal computer.

The data is read in column-wise, all rows in parallel; 20 bidirectional I/O pins are provided, one per row of cells. Therefore the on-chip digital I/O control block contains a shift register of Master-Slave flip-flops which select one column after the other to transfer the sensor signals. The shifting of the SET pulse to form the different COL_SEL signals is represented in figure 4.30(b). The flip-flops have been designed as static CMOS S-R flip-flops; the complementary set signals and clocking signals are derived on chip from the external references.

Bidirectional I/O pin circuitry. In figure 4.30(a) the schematic of the control and driver circuitry of the bidirectional row I/O pins is represented. When the OUT/IN control signal is low, the input data is read into the chip. For a high OUT/IN signal, the output mode is selected; an analog output value is provided with the buffer/comparator circuit if the COMP/FOL signal is kept low; then the buffer is put in a unity feedback mode and its buffers the large external pin and load capacitance; this mode can be used to monitor the analog final states of the cells or in a special I/O mode to follow the analog evolution of a row of cells during computations. For a high COMP/FOL signal the buffer is put in a comparator configuration and outputs a digital high for a black pixel and a digital low for a white pixel.

All cells are connected to the bidirectional pin buffer/comparator driver via the common ROW_IO line, which runs across a row of cells; it has a relatively large parasitic capacitance and to avoid any crosstalk between the output signals from the previous read cell to the presently read cell, a pre-charge of the ROW_IO line is executed in between the reading of different cell outputs. This pre-charge is controlled with the PRECH signal.

In the present set-up all these control signals are generated by the global system controller i.e. the PC. These control signals are all buffered in the control block of the chip and the necessary complementary signals are also derived on chip. The required sequences of control signals have been represented in figure 4.30(b) for a complete input-computation-output sequence of a 5x5 network; the signals generated by the global controller and the internal COL_SEL signals of the APAP I/O system are shown together with the required timing of the presentation of the input-data and the valid output-data sample periods.

The I/O system has been designed and its correct operation has been simulated for a clock frequency up to 250 kHz with an input accuracy of better than 6 bits; the minimal required time to input or output a frame is only 80 μs. The output buffer/comparator has a load-compensated topology and its stability is guaranteed even without output loading; the buffer can drive a load up to 20 pF and the calculated standard deviation of its input referred offset voltage is 1.4 mV.

4.7.5 Test-set-up

A test and demonstration system for real-time sensor signal processing has been developed; in figure 4.31 a photograph of the set-up is presented; the set-up consists of different parts:

- a printed circuit board (PCB) containing the APAP chip;

- a personal computer (PC) with dedicated software to control the system;

- and different types 2D array sensors and interfaces.

APAP PCB. The APAP chip is the computing core of the system; it is mounted on an PCB containing interface circuitry towards the PC and the sensors. This board contains support circuitry like D/A converters for the weight tuning system, and A/D converters and current measuring circuits for the automatic measurement of cell characteristics. Also the necessary computer controller multiplexers and switches are provided to enable a selection of the

VLSI IMPLEMENTATION OF CNN'S   183

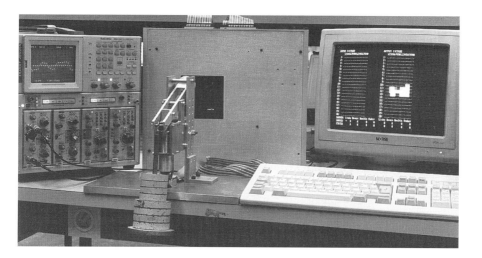

Figure 4.31. Photograph of the test- and demonstration set-up for the APAP chip and the real-time sensor signal processing system; the tactile sensor is located in the center and the APAP chip and PCB is located behind the sensor; the PC controls the operation and displays the processed signals which can also be monitored with the oscilloscope on the left.

sensor or the computer as the APAP input source; through this interface computer generated input images are presented to the chip to evaluate performance under worst case situations. Extra I/O circuitry to enable the connection of the APAP to a switch-matrix and digital sampling oscilloscope are also implemented so that the individual state evolutions of cells can be followed, sampled and stored (see e.g. fig. 4.34(b)).

Controller. The PC acts as the global controller; it contains a 96-lines digital I/O interface card to generate the control signals for the APAP chip and the APAP support-PCB and to read the processed frame from the APAP chip. The controller function of the PC is implemented in a dedicated software package; it contains a library of high-speed C routines for the low-level interfacing to the I/O interface and the APAP board and a MS Visual Basic software that provides the graphical user interface. The interface screen of the software is presented in figure 4.34(a). The user selects the computer or the sensor as input source; for a computer generated input, he can assemble an input image on the left side of the screen. Then the template settings are selected and the control signals are loaded from a configuration file; at this point the image can be processed by pushing the "RUN" button. For the processing of sensor images, also a continuous operation mode can be selected so that the output signal screen continuously presents the processed sensor image. For testing purposes also interfaces towards the HP-IB workstation controllers have been written for automatic measurements and for higher speed testing standard MS-DOS controller routines are available.

2D array sensors. Different types of sensors are available to demonstrate and test the sensor signal processing system; basically any 2D array sensor with a column-wise parallel row output of analog voltages can be connected to the system. Detailed tests with two type of sensors have been performed: a tactile sensor for robot arm control applications and a finger position sensor.

Tactile Sensor. The mechanical part of the sensor consists of a pattern of metal patches connected in rows and column tracks etched on a two-sided PCB board and covered with a layer of conductive rubber as pressure transducing material; the resistance of the rubber is proportional to the local pressure exerted by the object pressed on it [Reyn93].

The interface electronics of this sensor have been redesigned to allow for a high speed parallel operation; a voltage is applied to one column at a time and the other columns are connected to ground; all rows are kept at ground potential by connecting them to trans-impedance amplifiers [Lak 94], with a virtual ground input; the current flowing through the rubber from a column track to one row patch is measured; the parasitic current paths through the other patches connected to the row track are all shorted to ground and carry no current. In this way the sensor has a column-wise parallel row output. The analog output voltages are proportional to the current through the rubber for a fixed voltage drop and thus proportional to the local pressure.

Finger position sensor. Using a similar PCB a finger position sensor has been made. In this case a tiny current is flowing through the resistance of the human skin so that at the position of the finger output signals are generated. The same trans-impedance based electronic interface is used to generate column-wise, parallel row output signals.

4.7.6 Chip and Measurement Results

4.7.6.1 Lay-out. The APAP chip is laid out and fabricated in a standard digital 0.7 μm CMOS technology with two metal levels. In figure 4.32 a microphotograph of the chip is represented. The active area of the chip is 4.9 mm by 4.9 mm. A single cell of the matrix measures 245 μm by 245 μm which corresponds to a cell density of 16.7 cells/mm^2. Only poly-silicon and metal 1 is used for the intra-cell connections. The global routing over the cells is in metal 2. The input and state capacitors are put under the routing.

The digital control part of the I/O circuits is located at the top of the die; the bidirectional I/O pins with their driving circuits are located on the right side of the die; all these circuits are very compact and require almost no extra space on the die. The weight tuning circuits are simply put on empty places between the bonding pads.

4.7.6.2 Functional testing. In figure 4.33(a) the measurements of the output current of an A-template circuit as a function of the state voltage is shown for different weight control voltages. The attained range of template values covers the 1/4 to 4 range as can be evaluated from figure 4.33(b), where the A-template weights are plotted as a function of the weight control voltage $V_{CONTROL}$.

The chips were first extensively tested separately for a large set of template settings with computer generated test patterns. In figure 4.34(a), for instance, a test input image and the corresponding output image is represented for the connected component detector to the right. With these test-images faulty cells or stuck at faults can be detected.

All 15 samples of the APAP chip we received from the foundry showed satisfactory functional operation; only one chip contained a cell with a catastrophic

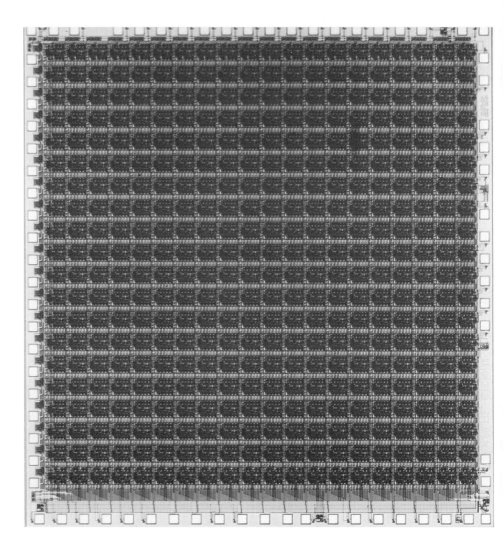

Figure 4.32. Micro-photograph of the APAP chip containing 400 computing cells, I/O, template and control circuits.

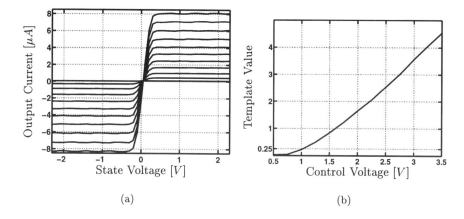

Figure 4.33. (a) Measured output current of an A-template circuit as a function of the state voltage for different weight control voltages. (b) Measured A template value as a function of the weight control voltage.

error due to which it remained stuck-at-white; all other cells of this chip functioned correctly.

4.7.6.3 Parametric testing.

Speed. Using computer generated test patterns the chips are tested parametrically under worst-case conditions. In figure 4.34, for instance, the analog state evolution of a column of cells of the connected component detector to the right is shown for the input pattern represented in figure 4.34(a). The outermost left cell is initialized high (black) and the other cells are initialized low (white). The single black connected component is shifted from cell to cell to the right and stops at the edge. The last cell has a high final state and thus a black output. All other cells have a white output.

In figure 4.34 the connected components are calculated in 175 μs. In order to measure the evolution of the cells, one row of cells is connected to the row I/O buffers through the ROW_IO line; this line represents an extra load on the internal state nodes so that the computation becomes slightly slower. In normal operation where the measured execution time is less than 145 μs. This corresponds to a cell time constant of of about 4.8 μs. For templates with information propagation like the connected component detector the execution time is proportional to N for an NxN network, whereas for templates without information propagation, like for instance noise filtering or edge detection, the

188 ANALOG VLSI INTEGRATION OF MASSIVE PARALLEL SYSTEMS

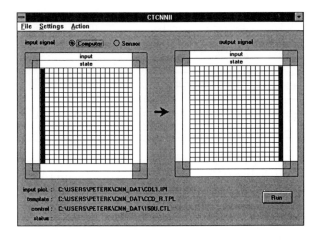

(a)

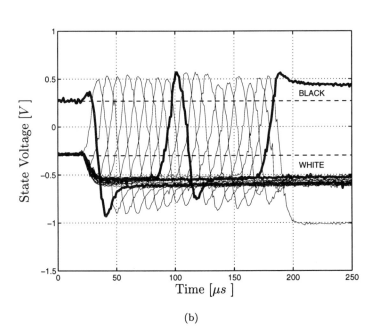

(b)

Figure 4.34. (a) Test Input picture and resulting Output picture for the connected component detector template. (b) Evolution of the state of a row of cells. The evolution of the first, the tenth and the last cell are emphasized. The first starts black (high) and the other cells start white (low); the only black connected component shifts from cell to cell to the right and stops at the edge.

execution time is *independent* of the network size. It typically takes 9.6 μs to execute a non-propagating template. The experimental derived value of the time constant deviates from the designed cell time constant ($R_{cell} \cdot C_{state}$) of 2.7 μs. This deviation is probably due to process variations in the absolute device parameters and in extra loading on the state node by the building blocks parasitics connected to the state node. The state node of the individual cells could not be reached on-chip since they are always buffered by the row buffer (see figure 4.30(a)) so that the exact cause of the deviation could not be tracked.

The I/O circuits can be clocked at 250 kHz so that 80 μs are required for the transfer of the sensor output to the chip or the APAP output to the PC. For a typical frame rate of 25 frames/sec, up to 3200 local instructions (i.e. non-propagating templates like e.g. noise filtering and edge detection) and over 200 global instructions (i.e. propagating templates like e.g. shadow making, hole-filling and connected component detection) can be executed per frame in real-time. This means that very complex image processing algorithms using the templates as elementary instructions can be executed by this system in real-time (see also figure 4.35).

Power Consumption. The power consumption of the programmable chip is dependent on the template settings. The DC power consumption of the state-to-output converter, the input buffer and the current buffer is fixed since they are all current-biased. The DC current consumption of the template circuits depends on the selected template value; the total DC current through the template circuits is given by:

$$I_{DC} = \left(3.8 \cdot \sum_i |A_i| + 3.8 \cdot \sum_i \cdot |B_i| + |I|/2\right) \cdot I_{UNIT} \qquad (4.42)$$

where the 3.8 factor is determined by the choice of the signal amplitudes and common mode biases of the template multipliers; this factor is slightly dependent on the selected weight value but 3.8 is the worst case value. For the connected component template 75 μA are consumed by a cell with a $+2.5/-2.5\ V$ power supply. For the holefiller template 150 μA are consumed by a cell which corresponds to a total consumption for the chip of 300 mW.

Accuracy. Due to the variation in the template values a statistical distribution of the final state values of the cells exists. This means that the accuracy of the template circuits, and thus also the circuit yield, can be checked by observing the statistical distribution of the cell final state x^*. The relative statistical

Table 4.8. Measured accuracy of the final states of the first 18 cells in a horizontal connected component detector tested with the input image of figure 4.34(a); the predicted value of the final states is 3.8%; the 95% confidence intervals on the standard deviation measurements for the sample size of 144 are also indicated.

	ACCURACY FINAL STATES CoCoD. (144 cells)		
chip	rel. acc. [%]	95 % Confidence interval min. [%]	max. [%]
3	3.72	3.34	4.21
8	3.99	3.58	4.51
14	3.99	3.58	4.51

variation of a cell final state x^* is given by:

$$\left(\frac{\sigma(x^*)}{x^*}\right) = \left(\frac{\sigma(G_{cell})}{G_{cell}}\right)^2 + \frac{\sum_i \sigma^2(A_i) + \sum_i \sigma^2(B_i) + \sigma^2(I)}{(\sum_i A_i + \sum_i B_i + I)^2} \quad (4.43)$$

For the connected component detector e.g. the statistical variation of the final states of cells with two white neighbors is 3.8% when the accuracy of the circuits of the APAP chip are substituted in equation (4.43).

The standard deviation of the final states of the cells of a connected component detector are determined from the type of measurement represented in figure 4.34(b). The last cell has a black final state. The 19th cell has a black right neighbor and a white left neighbor so that its final state is lower than the final state of the first 18 cells who have two white neighbors. The theoretical final state of the first 18 cells is $-2 \cdot V_{UNIT}$ or -0.6V. In table 4.8 the accuracy measurements of the final states are summarized for three chips. All measurements agree very well within the confidence of the measurements with the theoretical prediction. This implies that the template circuits achieve their accuracy specification so that for high volume productions a high parametric chip yield is guaranteed.

Real-time sensor interfacing. In figure 4.35 the experimental results of pressing a tube on the tactile sensor and applying a set of templates consecutively are shown. The final pictures contain different features of the object for intelligent robot control or object recognition, like for instance the x and y position of the center of the object, which is used for real-time slip detection [Reyn93].

VLSI IMPLEMENTATION OF CNN'S 191

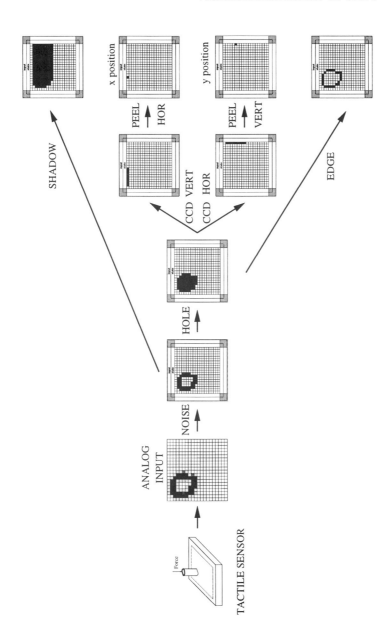

Figure 4.35. Measured results of template sequences (from table 1.2) executed on the tactile sensor output.

Table 4.9. Summary of the experimental performances of the 20x20 analog parallel array processor chip.

	CELL DENSITY	
0.7 μm CMOS	16.7 cells/mm^2	
	SPEED	
Cell time constant	4.8 μs	
I/O of frame	80 μs	
Real time operation	212 global operations	
(25 frames/sec)	3200 local operations	
	POWER CONSUMPTION	
Connected Component detector	75 μA/cell	(375 μW)
Holefiller	150 μA/cell	(750 μW)
	PROGRAMMABILITY	
A & B template weight values	±1/4 to ±4	

4.7.7 Evaluation

In table 4.9 the performance of the APAP chip and the sensor signal processing system is summarized. We can conclude that the main contribution of this chip is its design methodology which takes fully into account the accuracy specifications that are derived in chapter 3 together with the quantitative technology data (see chapter 2). This has resulted in a very robust design; all received samples operate satisfactory and the template circuits achieve the set accuracy specifications.

4.8 PERFORMANCE EVALUATION WITH OTHER IMPLEMENTATIONS

4.8.1 Comparison with other analog CNN implementations

In table 4.10 the performances and characteristics[†] of other analog continuous-time CNN implementations are summarized. The first two columns contain fixed template CNN implementations. The third column is a 1D discretely

---

[†]The connections are labeled N: north, E: east, S: south, and so on ...

Table 4.10. Overview of the published performance of experimentally verified CNN chips with fixed, discretely or continuously programmable functionality.

	Cruz [Cru 91]	Espejo [Esp 94b]	Paaiso [Paa 94]	Espejo [Esp 96]	4x4 CNN [Kin 95]	APAP [Kin 96c]
CMOS Technology	2 μm	1.6 μm	1.2 μm	1.0 μm	2.4 μm	0.7 μm
Network size	6x6	16x16	1x5	32x32	4x4	20x20
Connections	N-S	N-S	N-S	N-E-S-W NE-SE SW-NW	N-E-S-W	N-E-S-W
Cell density (cells/mm^2)	31	88	8	31	3	16.7
Speed Cell τ (μs)	0.2	0.14	0.5	1-2(?)	10	5
Power (mW/cell)	?	0.105	?	?	0.5 (CoCoD)	0.375 (CoCoD)
Programmability	fixed CoCoD	fixed CoCoD	discrete	contin. ?	contin. ±1/4-±4	contin. ±1/4-±4

programmable CNN implementation whereas the last three columns are continuously programmable CNN implementations, containing the two chip implementations presented in this work.

Fixed Function Chips. The cell density of fixed function chips and their speed and power consumption are considerably better than the performances of the programmable implementations. This illustrates again that programmable template circuits are more area and power consuming since they have to be designed for worst case conditions, whereas the fixed function circuits can be optimized for their single function. This is further illustrated by the numbers in table 4.11. If we compare the individual obtained accuracies for the template implementations with their required values, we see that in a programmable circuit too high values are obtained. This is a direct consequence of the necessary

Table 4.11. The allowed $\sigma(\Delta_{route})$ for a yield of 90 % of a 20x20 network and the $\sigma(\Delta_{route})$ obtained in the APAP chip realization.

	SPECIFICATION		REALIZATION
Template	Δ_{max}	$\sigma(\Delta_{route})$	$\sigma(\Delta_{route})$
CoCoD	1	0.27	0.07
HOLE_MOD	1	0.27	0.11
SHADOW	2	0.55	0.09
EDGE	1	0.27	0.11
NOISE	1	0.27	0.09
PEEL	3	0.82	0.10

worst case design necessary in programmable implementations and this is the fundamental reason why programmable designs will always be less performant than fixed function implementations.

Discrete programmable Networks. The cell density of discretely programmable networks, reported in the third column of table 4.10, is worse than for the circuits using analog multipliers especially if we take into account that in a 1D network only three A and three B template per cell multipliers are necessary. These chip results confirm the statement made in section 4.3.2C that the area consumption due to interconnection overhead in discretely programmable networks is very large.

Analog Programmable Networks. At present only two large fully analog programmable CNN implementation have been published in open literature [Kin 96c, Esp 96]. The analog complexity and the analog functionality of the implementation of [Esp 96] and the 20x20 APAP [Kin 96c] are comparable. The [Esp 96] chip includes extra digital functions to perform elementary logic operations in the cells on the results of the analog computations and local memory to store a few values. However the B template circuits are not physically implemented; the constant contribution of the B template is pre-computed using the A template circuitry in an extra computation step; this results in important area savings but requires more complex timing and slows down the computations.

The cell time constant of both implementations is in the same range but cannot be compared since the power consumption is not reported included in [Esp 96]. Moreover, in [Esp 96] a full range CNN cell model [Esp 94a] is implemented which only has one time constant in the circuit present; this represents about a 10 times speed advantage due to the second order time constants present at the output node in the standard CNN model.

In [Esp 96] analog template multipliers of the type in figure 4.8(c) are used with the weight value signals connected to the drains; these weight signals must be buffered by very low output impedance buffers which are very power hungry. Finally, the results in [Esp 96] do not include measurement results of a fully functional prototype so that these results must be considered preliminary.

We can conclude that the analog programmable circuit implementations presented in this work have a state of the art performance; the analog multiplier circuit solutions are also more suitable for the realization of programmable CNN's then discrete programmable multiplier implementations.

4.9 CONCLUSIONS

In this chapter the VLSI implementation of massively parallel analog signal processing systems is discussed and the design of fully programmable CNN systems for sensor signal processing is used for the practical chip implementations.

For the different computational operations, different physical signal representations are used. Current signals can easily be added; current signals can also be integrated on a capacitor and the voltage across the capacitor represents the integral. The programmable weighting or multiplication of signals is an operation that is extensively used in parallel signal processing systems. The implementation of compact analog multipliers in standard digital CMOS technologies is however a challenging design problem. We have shown that multipliers based on the operation of the MOS transistor in the linear region are best suited for this application. Two analog multiplier implementations are introduced: a programmable current mirror circuit and a compact linear grounded transconductor multiplier circuit.

The design of an efficient input/output strategy and circuits for analog parallel signal processing systems is also essential to obtain a high density system chip design. Several combinations of the processing hardware and the sensor hardware have been evaluated; for the design of fully programmable processing hardware, a design with a separate sensor (if possible on the same die) and an X-Y addressing based interface is shown to be the best input/output solution.

These basic circuits have then be used to assemble a VLSI cell architecture for the implementation of fully programmable analog CNN chips. Besides compact circuits for the different building blocks, also a bias generation and weight tuning strategy has been used. This allows a very flexible use of the chips and the tuning loops compensate many of the systematic errors or process variations so that a very compact cell circuit implementation is obtained.

Two chip realizations of this architecture have been presented. A 4x4 fully programmable CNN prototype chip has clearly demonstrated the feasibility of the realization of programmable CNN's in standard digital CMOS technologies. The second chip realization is a 20x20 fully programmable CNN which is used as an analog parallel array processor for real-time sensor signal processing. This chip has been designed and sized with a design methodology that takes into account the accuracy specifications derived in chapter 3 and that uses the quantitative mismatch model data of chapter 2 for the sizing of the circuits towards a maximal cell density; this design methodology guarantees a high chip yield. A real-time sensor signal processing test-set-up, including a tactile sensor and a finger position sensor and controlled by a personal computer, has been built. The experimental results demonstrate the real-time performance of the system; moreover, the obtained experimental chip yield is very high. Compared to other analog CNN implementations, these experimental results show the state-of-the-art performance of the presented devices.

5 GENERAL CONCLUSIONS

Biological information processing systems show a remarkable performance in real-time sensory data processing, in perception tasks, and in motory control applications; their performance is far beyond the capabilities of present-day artificial information processing systems. Artificial neural networks imitate the biological systems and use a similar massively parallel architecture of simple processors. In this work we have discussed the analog VLSI implementation of massively parallel signal processing systems. The continuous increase in the integration density of (digital) CMOS VLSI technologies and the advances in the design of compatible analog circuits in CMOS provide the fabrication tools for the realization of these systems. For the near future, these analog parallel signal processing systems show very promising applications in the field of real-time 2D (two-dimensional) sensor signal processing and perception-like applications.

The design of circuit implementations of ANN (artificial neural networks) requires a combined research activity on two levels. On the system level, the influence of circuit inaccuracies and non-idealities on the correct system behavior must be computed so that detailed specifications for the circuit building blocks can be generated. This implementation-oriented theory provides a toolset to the circuit designer to develop compact realizations and to optimize the VLSI implementation. In standard digital CMOS technologies, especially the design of multiplication circuits to build programmable systems is a challenging problem.

The circuit realization should attain a high integration density, high operation speed combined with a low power consumption. To increase the integration density, small device sizes are used but this increases the variation of the circuit properties. To increase signal processing accuracy, large device sizes are used but the higher capacitive loading of the circuit nodes then results in higher power consumption to achieve the speed requirements. Device mismatch is thus a major technological factor in the circuit design for analog parallel signal processing systems.

In this work, we have studied the technology aspects, the theoretical aspects and the circuit design aspects of the analog VLSI implementation of massively parallel signal processing systems and combined the research results to build a real-time 2D sensor signal processing system. As a practical demonstrator for this combined development strategy, programmable cellular neural networks have been used, but many of the results and conclusions are valid or can be extended for different other types of adaptive or massively parallel signal processing systems.

The implications of transistor mismatch on the design and performance of analog circuits and systems is covered in chapter 2. To provide a solid quantitative basis for the analysis, a characterization method has been developed and quantitative mismatch model data has been experimentally collected. The dependence of parameter mismatches on the device dimensions and position is modeled accurately for sub-micron CMOS technologies.

In elementary voltage or current processing circuits the total performance ratio $Gain^2 \cdot Speed \cdot Accuracy^2/Power$ is determined by the chosen biasing point or gate-overdrive voltage $(V_{GS} - V_T)$ and the technology matching quality. Current stages are optimally biased with large $(V_{GS} - V_T)$ whereas voltage stages attain the best total performance for a low $(V_{GS} - V_T)$. For these bias points the circuits attain the best *combined* performance for Gain, Speed, Accuracy and Power consumption. For the design of complex analog circuits and systems, including multiple stages and feedback, device mismatch again imposes a minimal amount of power for a certain speed or frequency performance and a certain accuracy performance and it limits the maximal attainable *Speed·Accuracy²/Power* ratio. This result explicitely proves that a designer can only trade one specification for another, and technological limitations restrict the ultimate total performance that can be achieved.

GENERAL CONCLUSIONS

The matching quality of the technology is inversely proportional to its[†] $C_{ox}A_{VT0}^2$ and we have shown that the impact of mismatch on the minimal power consumption is orders of magnitude more important than the physical limitation imposed by thermal noise for high speed and massively parallel analog systems. This equally applies for all other analog systems where no offset compensation or calibration strategies can be included. The analysis of the scaling of the mismatch behavior with the technology feature size forecasts an improvement of the matching quality for deeper sub-micron technologies. For very deep-sub-micron technologies, however, the presently available experimental data indicates that further downscaling will not or only marginally improve the analog performance[‡]. Due to the importance of transistor mismatch for the design of many analog systems good matching models for the used technology should always be available to the analog designer.

The main objectives for the circuit implementation of massively parallel analog signal processing and computational systems are a high circuit density, a high operation speed and a low power consumption. In chapter 2 the coupling between these specifications due to the effect of transistor mismatches is shown. Consequently accuracy specifications are indispensable for obtaining a high performance chip realization. Most theoretical descriptions and definitions of ANN do not contain enough information for the circuit designer to evaluate the effect of circuit non-idealities on the correct system operation. In chapter 3, we present the necessary implementation-oriented theory for the design of programmable cellular neural networks (CNN's). The effect of weight variations due to random errors on the yield of the CNN-chips is calculated using a new evaluation method and accuracy specifications are generated for the design of CNN chips including fully programmable CNN chips. The effect of random dynamic errors or mismatches in the time constants is shown to be insignificant for large classes of templates. The impact of systematic static errors has been studied in detail and we show that they have no significant influence on the circuit design. Parasitic time constants in the cell realizations cause systematic dynamic errors and must be kept about 10x smaller than the dominant time constant to avoid faulty cell operation. These new methods and results allow the circuit designer to explicitely determine the necessary specifications for the different circuit blocks. Moreover, they allow to optimize the parameters of the circuit implementation while guarantee-ing a correct system operation and a high yield for the fabricated chips. The evaluation methods have also been applied for the design and optimization of robust templates resulting in more relaxed circuit requirements and important area, speed or power consumption savings.

[†]C_{ox} is the gate oxide capacitance per unit area and A_{VT0} is the proportionality constant between the standard deviation of the random threshold voltage (V_T) difference off two transistors and $1/\sqrt{\text{gate area}}$

[‡]The behavior is predicted for technologies with a feature sizes around 0.25 μm and smaller.

Chapter 4 concentrates on the circuit design aspects of programmable massively parallel systems and the experimental results for two programmable CNN systems are reported. First the compact VLSI circuit implementation of the different computational operations is discussed. In standard digital CMOS technologies the design of programmable scaling circuits or analog multipliers with a large range in the scale factors requires extra attention. We present new compact multiplier circuits that are based the linear operation region of an MOS transistor. They have the best performance for massively parallel analog computational systems compared to other multiplier circuit realizations. An efficient input/output strategy is also essential for a successfull hardware implementation. For programmable systems we have developed an efficient X-Y cell addressing scheme which provides an interface to separate on-chip 2D sensors or external off-chip 2D sensors.

The basic circuits are combined into a cell circuit architecture for the implementation of fully programmable CNN chips. A bias generation and weight tuning strategy has been developed; they provide an easy use of the chips and compensate many of the systematic errors or process variations so that a more compact circuit implementation is achieved. Two chip realizations of this architecture are presented. A 4x4 fully programmable CNN prototype chip in 2.4 μm CMOS provided the first experimental proof in open literature of the feasibility of the realization of programmable CNN's in standard digital CMOS technologies. The second realization is a 20x20 fully programmable CNN in 0.7 μm CMOS which is applied as an analog parallel array processor for real-time 2D sensor signal processing. For the design of this chip the technological and optimal circuit design information of chapter 2 is combined with the theoretical evaluation methods of chapter 3 and the basic computational circuits and cell circuit architecture developed in the first part of chapter 4. This has resulted in a constrained circuit optimization design methodology which guarantees a high cell density combined with a high chip yield. A real-time sensor signal processing test-set-up has been built and it includes a tactile sensor and a finger position sensor and is controlled by a personal computer. The experimental results demonstrate the real-time performance of the system; moreover, the obtained experimental chip yield is high. The cell density is 16.7 cells/mm^2 and for the execution of the connected component template 375 μW is consumed per cell. For a rate of 25 frames/sec, 3200 local or 212 global operations can be executed in real-time. This represents a state-of-the art performance for analog fully programmable CNN implementations.

We conclude this book with a few indications for future work. The presented results demonstrate the feasability of realizing massively parallel analog signal processing systems for 2D sensor signal processing. Other research groups have also reported remarkable results in this area [Mea 89, Kob 90, Sta 91, Gru

91, Yu 92, Esp 94b, Che 95, Arr 96, Veni96]. These developments will support the design and fabrication of new types of smart sensors and sensor interfaces which include signal conditioning, feature extraction and A/D conversion functions. With a further down-scaling of technology it will be possible to integrate larger and larger systems on a single die. However, the interconnection of multiple chips to build very large arrays of cells is also an interesting alternative to increase the system capabilities and performance. This will require the development of an efficient communication channel between the chips so that a form of output and input pin multiplexing between many cells can be achieved.

On the system level, two activities could significantly improve the application of massively parallel analog systems. Many of the practical limitations for the circuit implementations have now been studied; these results should be used to redefine and to optimize existing systems or to develop new systems which are adapted to an analog VLSI implementation. Secondly, given the assets and the constraints of this technology more application areas should be identified where the advantages of these systems can be exploited.

On the hardware level, a further optimization of the circuits and topologies will be required for the newer deeper sub-micron technologies; especially the lower supply voltages and more pronounced short and narrow channel effects will force the designers to come up with new solutions for certain functions and will maybe degrade certain specifications. On the other hand, the availability of higher intrinsic transistor speeds, better intrinsic device matching and more wiring levels should enable an improvement of the performance. For the design of ANN's including adaptivity and learning capabilities, the design of a compact analog memory cell is a very important research issue. This will probably involve both circuit design issues as technological modifications.

The results presented on the fundamental impact of device mismatch on the performance of many analog systems, should encourage analog designers to take mismatch carefully into account. They should also prompt foundries to carefully monitor the matching quality of their technology and to provide designers with quantitative device mismatch models. For the development of new technologies the evolution of the matching of devices should be taken into account.

Appendix A
MOS transistor models

In this appendix we summarize a few MOS transistor model equations that are used throughout the text. These equations are mainly first order approximations that can be used for hand calculations; more accurate model equations can be found in [Lak 94] and [Tsi 88]. The spice model cards for the used technologies are included at the end of this appendix.

The used biasing conventions are summarized in figure A.1(a) and A.1(b) and the small signal equivalent model is shown in figure A.1(c). The figures use an nMOS transistor; for pMOS transistors equivalent conventions are used.

Definition of symbols

β $KP \cdot \frac{W}{L}$; for large vertical or lateral electrical fields in strong inversion the mobility degradation has to be taken into account and $\beta = KP \cdot (W/L)/(1 + \theta(V_{GS} - V_T) + \mu V_{DS}/(V_{max}L))$, where V_{DS} is to be replaced by $(V_{GS} - V_T)$ for transistors operating in saturation (spice parameters: $\theta =$ THETA $[1/V]$, $\mu =$ U0 $[cm^2/(Vs)]$ and $V_{max} =$ VMAX $[cm/s]$);

λ $1/(V_E \cdot L)$, where V_E $[V/m]$ is the early voltage of the technology;

γ Bulk threshold parameter (spice paramter GAMMA)

C_{ov} overlap capacitance between gate and source or drain (spice parameter CGSO and CGDO $[F/m]$);

C_{ox} $\frac{\epsilon_{ox}}{t_{dx}}$; $\epsilon_{ox} = 33.8 \ pF/m$

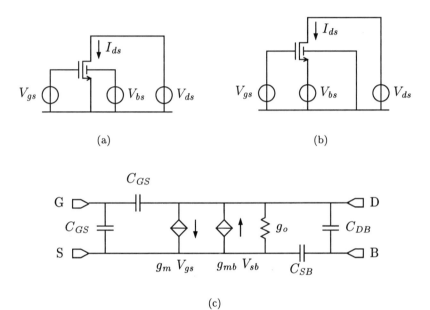

Figure A.1. Biasing conventions with respect to the source (a) or bulk (b) terminal. (c) The small signal equivalent schematic for a MOS transistor.

APPENDIX A: MOS TRANSISTOR MODELS   205

I_{D0} specific current in weak inversion, exponentionaly dependent on v_{BS} and V_T;

k Boltzmann constant $1.38 \cdot 10^{-23}$ J/K;

n subthreshold slope: $n = 1 + \frac{C_{BC}}{C_{GC}}$, where C_{BC} is the bulk-channel capacitance and C_{GC} is the gate-channel capacitance;

KP $\mu \cdot C_{ox}$ current factor parameter (spice parameter: $\mu =$ U0 [cm^2/(Vs)]);

L the length of the transistor gate;

T absolute temperature in Kelvin;

t_{ox} gate oxide thickness (spice parameter TOX [m]);

U_T $\frac{k \cdot T}{q} = 25.8\ mV$ at room temperature;

V_T threshold voltage;

W the width of the transistor gate;

Weak Inversion

An MOS transistor is operating in weak inversion if:

$$v_{GB} < v_{SB} + V_T \quad (A.1)$$
$$\text{or} \quad v_{GS} < V_T \quad (A.2)$$

The saturation voltage is very small and equal to a few U_T; consequently the transistor is operating in saturation in most practical cases.

Saturation DC current:

$$i_{DS} = I_{D0} \cdot \left(\frac{W}{L}\right) \cdot \exp\left(\frac{v_{GS}}{n \cdot U_T}\right) \quad (A.3)$$

AC small signal parameters:

$$g_m = \frac{\partial i_{DS}}{\partial v_{GS}} = \frac{I_{DS}}{n \cdot U_T} \quad (A.4)$$

$$g_{mb} = \frac{\partial i_{DS}}{\partial v_{SB}} = (n-1) \cdot g_m \quad (A.5)$$

$$g_o = \frac{\partial i_{DS}}{\partial v_{DS}} = \lambda \cdot I_{DS} \quad (A.6)$$

Intrinsic small signal capacitances:

$$C_{GS} = C_{GB} = \frac{n-1}{n} \cdot C_{ox} W \cdot L + C_{ov} \cdot W \quad (A.7)$$

Strong Inversion

A MOS transistor is operating in strong inversion if:

$$v_{GB} \geq v_{SB} + V_T + \Phi_B \quad \text{(A.8)}$$
$$\text{or} \quad v_{GS} \geq V_T + \Phi_B \quad \text{(A.9)}$$

with $\Phi_B \geq 150$ to $200\ mV$; for smaller values of $(V_{GS} - V_T)$ (> 0) the drain-source current contains a considerable contribution due to diffusion current and the transistor is in moderate inversion [Tsi 88].

Linear/Triode Region Bias condition:

$$v_{DS} \leq (v_{GS} - V_T) \quad \text{(A.10)}$$

DC current:

$$i_{DS} = \beta\,(v_{GS} - V_T - v_{DS}/2) \cdot v_{DS} \quad \text{(A.11)}$$
$$i_{DS} = \beta\,(v_{GB} - V_T - (v_{DB} + v_{SB})/2) \cdot (v_{DB} - v_{SB}) \quad \text{(A.12)}$$

Approximation for $v_{DS} \ll (v_{GS} - V_T)$:

$$i_{DS} \approx \beta\,(v_{GS} - V_T) \cdot v_{DS} \quad \text{(A.13)}$$

AC small signal parameters:

$$g_m = \frac{\partial i_{DS}}{\partial v_{GS}} = \beta \cdot V_{DS} \quad \text{(A.14)}$$

$$g_{mb} = \frac{\partial i_{DS}}{\partial v_{SB}} = (n-1) \cdot g_m \quad \text{(A.15)}$$

$$g_o = g_{DS} = \frac{\partial i_{DS}}{\partial v_{DS}} = \beta \cdot (V_{GS} - V_T - V_{DS}) \quad \text{(A.16)}$$

Intrinsic small signal capacitances:

$$C_{GS} = 1/2 \cdot C_{ox} W \cdot L + C_{ov} \cdot W \quad \text{(A.17)}$$
$$C_{GD} = 1/2 \cdot C_{ox} W \cdot L + C_{ov} \cdot W \quad \text{(A.18)}$$

Saturation Region Bias Condition

$$v_{DS} > (v_{GS} - V_T) \quad \text{(A.19)}$$

APPENDIX A: MOS TRANSISTOR MODELS

DC current:
$$i_{DS} = \frac{\beta}{2} \cdot (v_{GS} - V_T)^2 \cdot (1 + \lambda \cdot v_{DS}) \tag{A.20}$$

AC small signal parameters:
$$g_m = \frac{\partial i_{DS}}{\partial v_{GS}} = \beta \cdot (V_{GS} - V_T) \tag{A.21}$$

$$= \sqrt{2 \cdot \beta \cdot I_{DS}} \approx \frac{2 \cdot I_{DS}}{(V_{GS} - V_T)} \tag{A.22}$$

$$g_{mb} = \frac{\partial i_{DS}}{\partial v_{SB}} = (n-1) \cdot g_m \tag{A.23}$$

$$g_o = g_{DS} = \frac{\partial i_{DS}}{\partial v_{DS}} \approx \lambda \cdot I_{DS} \tag{A.24}$$

Intrinsic small signal capacitances:
$$C_{GS} = 2/3 \cdot C_{ox} W \cdot L + C_{ov} \cdot W \tag{A.25}$$
$$C_{GD} = C_{ov} \cdot W \tag{A.26}$$
$$\tag{A.27}$$

Extrinsic Parasitic Capacitances

The drain and source terminals form a diode with the bulk so that they also have a parasitic junction capacitance to the bulk. Strictly speaking the gate-drain and gate-source overlap capacitances are also extrinsic parasitic elements but for compactness we have already included them in the expressions of C_{GS} and C_{GD} for the different regions.

The S/D-bulk capacitances are given by:
$$C_{SB} = C_{je} W \cdot L_{ed} \tag{A.28}$$
$$C_{DB} = C_{je} W \cdot L_{ed} \tag{A.29}$$

when the side-wall capacitance is neglected; C_{je} is the effective junction capacitance and depends on the S/D-bulk bias voltages; L_{ed} is the length of the drain and source diffusion regions and is determined by the technology lay-out rules.

For circuit design it is convenient to express the S/D-bulk capacitances as function of the gate-source capacitance [Stey93]. For a minimal length transistor biased in strong inversion and in saturation we define:

$$\alpha_{DB} = \frac{C_{DB}}{C_{GS}} = \frac{3}{2} \cdot \frac{C_j}{C_{ox}} \cdot \frac{L_{ed}}{L_{min}} \tag{A.30}$$

which is determined by technology constants.

f_T and f_{max} of a transistor

The intrinsic speed of a transistor is often defined [Lak 94] as:

$$f_T = \frac{g_m}{2\pi C_{GS}} = \frac{3}{4\pi} \frac{\mu(V_{GS} - V_T)}{L^2} \qquad (A.31)$$

This value is maximal for a minimal length transistor and depends on the chosen bias point $(V_{GS} - V_T)$ for the transistor. The f_T gives a first indication of the maximal speed that can be attained in a technology for a given bias condition. In most circuit designs however, the bandwidth or maximal speed is degraded due to the extra loading of the transistor by the extrinsic parasitic capacitors. Therefore the f_{max} is defined as the cut-off frequency of a diode connected transistor [Stey93]:

$$f_{max} = \frac{g_m}{2\pi(C_{GS} + C_{DB})} \qquad (A.32)$$

$$= \frac{f_T}{1 + \alpha_{DB}} \qquad (A.33)$$

The f_{max} gives a more realistic indication of the maximal speed that can be attained in a given technology.

Technology model parameters

In this section the SPICE model decks for the used technologies are included. For the definition of the different model parameters the reader is referred to the definition of symbols and [Lak 94].

2.4 μm CMOS

NMOS

Spice level 3 parameters:
```
.model nmos3 nmos level=3 vto=0.86 nsub=1.39e15 tox=40.29n gamma=0.:
+ ld=0.22u kp=51.71u phi=0.62 uo=611.37 delta=0.85 cj=6.9e-5 mj=0.5
+ pb=0.65
+ mjsw=0.27 rsh=33.42 vmax=158e3 cjsw=3.43e-10 cgbo=5.57e-10
+ nfs=1.35e11 ucrit=10000 eta=0.07 kappa=1.4 xj=0.3u js=0.001 theta
+ capop=4 xw=0 xl=0 lmlt=1 wmlt=1 wd=-0.028u del=-0.16u hdif=3u acm
```

PMOS

Spice level 3 parameters:
```
.model pmos3 pmos level=3 vto=-0.85 nsub=9.1e15 tox=42.46n gamma=0.69
+ ld=0.35u kp=19.14u phi=0.67 uo=233.84 delta=0.96 cj=3.1e-4 mj=0.5
+ pb=0.76
+ mjsw=0.38 rsh=35.01 vmax=225.67e3 cjsw=3.67e-10 cgbo=5.57e-10
+ nfs=3.92e11 ucrit=10000 eta=0.06 kappa=9.23 xj=0.5u js=0.001 theta=0.12
+ capop=4 xw=0 xl=0 lmlt=1 wmlt=1 wd=0.034u del=-0.01u hdif=3u acm=2
```

0.7 μm CMOS

Spice level 2 and level 3 parameters are available; the foundry optimized the extraction of the level 2 parameters for analog circuit simulation, whereas the level 3 parameter extraction is optimized towards the simulation of digital circuits.

NMOS

Level 2 parameters:
```
.MODEL NA NMOS LEVEL=2
+ LMIN =3E-6 LMAX =4E-6 WMIN =2.0E-6 WMAX =1
+ VTO =750E-3 NSUB =7.0E16 DELTA=2.0 NFS =1.2E11
+ TOX =17E-9 UO =470 UCRIT=1.08E5 UEXP =0.124
+ LD =0.1E-6 DELL =0.2E-6 WD =0.12E-6 DELW =0
+ XJ =50E-9 CGSO =2.1E-10 CGDO =2.1E-10 RSH =480
+ JS =1E-3 PB =0.65 FC =0.5
+ CJ =5.0E-4 MJ =0.33 CJSW =2.8E-10 MJSW =0.214
+ ACM =3 HDIF =1.4E-6 LAMBDA=0.0085
+ KF =1E-28 AF =1 NLEV =0
```

In the above model card LAMBDA is specified for a transistor length $L = 3\ \mu m$; for transistors with a different length L, the value of LAMBDA is to be calculated with:

$$\text{LAMBDA} = (\text{LAMBDA}@L = 3\mu m) * \frac{(3 \cdot 10^{-6} - 2 * \text{LD} + \text{DELL})}{(L - 2 * \text{LD} + \text{DELL})} \quad (A.34)$$

Level 3 parameters:
```
.MODEL ND NMOS LEVEL=3
+ LMIN =0.7E-6 LMAX =1 WMIN =2.0E-6 WMAX =1
+ VTO =750E-3 NSUB =7.0E16 DELTA=1.0 NFS =1.2E11
+ KAPPA=1E-3 VMAX =1.94E5 ETA =5.2E-3
```

+ TOX =17E-9 UO =470 THETA=79.1E-3
+ LD =0.1E-6 DELL =0.2E-6 WD =0.12E-6 DELW =0
+ XJ =50E-9 CGSO =2.1E-10 CGDO =2.1E-10 RSH =480
+ JS =1E-3 PB =0.65 FC =0.5
+ CJ =5.0E-4 MJ =0.33 CJSW =2.8E-10 MJSW =0.214
+ ACM =3 HDIF =1.4E-6
+ KF =1E-28 AF =1 NLEV =0

PMOS

Level 2 parameters:
.MODEL PHA PMOS LEVEL=2
+ LMIN =3.0E-6 LMAX =4.0E-6 WMIN =2.0E-6 WMAX =1
+ VTO =-1.00 NSUB =3.5E16 DELTA=2.0 NFS =5.2E10
+ TOX =17E-9 UO =148 UCRIT=1.3E5 UEXP =0.252
+ LD =0.06E-6 DELL =0.15E-6 WD =0.2E-6 DELW =0
+ XJ =50E-9 CGSO =1.2E-10 CGDO =1.2E-10 RSH =1000
+ JS =1E-3 PB =0.78 FC =0.5
+ CJ =6.0E-4 MJ =0.468 CJSW =3.6E-10 MJSW =0.302
+ ACM =3 HDIF =1.4E-6 LAMBDA=0.0100
+ KF =4E-30 AF =1 NLEV =0

In the above model card LAMBDA is specified for a transistor length $L = 3\ \mu m$; for transistors with a different length L, the value of LAMBDA is to be calculated with:

$$\text{LAMBDA} = (\text{LAMBDA@L} = 3\mu\text{m}) * \frac{(3 \cdot 10^{-6} - 2 * \text{LD} + \text{DELL})}{(L - 2 * \text{LD} + \text{DELL})} \quad (A.35)$$

Level 3 parameters:
.MODEL PHD PMOS LEVEL=3
+ LMIN =0.7E-6 LMAX =1 WMIN =2.0E-6 WMAX =1
+ VTO =-1.00 NSUB =3.5E16 DELTA=1.0 NFS =5.2E10
+ KAPPA=1E-3 VMAX =7.2E5 ETA =30E-3
+ TOX =17E-9 UO =148 THETA=130.4E-3
+ LD =0.06E-6 DELL =0.15E-6 WD =0.2E-6 DELW =0
+ XJ =50E-9 CGSO =1.2E-10 CGDO =1.2E-10 RSH =1000
+ JS =1E-3 PB =0.78 FC =0.5
+ CJ =6.0E-4 MJ =0.468 CJSW =3.6E-10 MJSW =0.302
+ ACM =3 HDIF =1.4E-6
+ KF =4E-30 AF =1 ,NLEV =0

Bibliography

[Ack 96] B. Ackland and A. Dickinson, "Camera on a chip," in *Digest of Technical Papers IEEE International Solid-State Circuits Conference (ISSCC)*, pp. 22–25, February 1996.

[All 87] P. Allen and D. Holberg, *CMOS analog circuit design*. HRW Series in Electrical and Computer Engineering, Fort Worth (Fla): Holt, Rinehart and Winston, Inc., 1987.

[And 88] J. Anderson and E. E. Rosenfeld, *Neurocomputing: foundations of research*. Cambridge (MA): MIT Press, 1988.

[Arr 89] X. Arreguit, *Compatible Lateral Bipolar Transistors in CMOS technology: Model and Applications*. PhD thesis, Ecole Polytechnique Federale de Lausanne, Lausanne (Switzerland), 1989.

[Arr 96] X. Arreguit, F. van Schaik, F. Bauduin, M. Bidiville, and E. Raeber, "A CMOS motion detector system for pointing devices," in *Digest of Technical Papers IEEE International Solid-State Circuits Conference (ISSCC)*, pp. 97–99, February 1996.

[Bas 95] J. Bastos, M. Steyaert, R. Roovers, P. Kinget, W. Sansen, B. Graindourze, N. Pergoot, and E. Janssens, "Mismatch characterisation of small size MOS transistors," in *Proceedings of the IEEE International Conference on Microelectronic Test Structures*, pp. 271–276, March 1995.

[Bas 96a] J. Bastos, *Matching characterization of MOS transistors for precision analog design*. PhD thesis, Katholieke Universiteit Leuven, Leuven (Belgium), to be submitted in 1996.

[Bas 96b]   J. Bastos, M. Steyaert, B. Graindourze, and W. Sansen, "Influence of die attachment on MOS transistor matching," in *Proceedings of the IEEE International Conference on Microelectronic Test Structures*, pp. 27–31, March 1996.

[Bas 96c]   J. Bastos, M. Steyaert, B. Graindourze, and W. Sansen, "Matching of MOS transistors with different layout styles," in *Proceedings of the IEEE International Conference on Microelectronic Test Structures*, pp. 17–18, March 1996.

[Bra 80]   R. Brayton and R. Spence, *Sensitivity and Optimization*. CAD of Electronic Circuits, 2, Amsterdam: Elsevier, 1980.

[Bul 87]   K. Bult and H. Wallinga, "A class of analog CMOS circuits based on the square-law characteristic of a MOS transistor in saturation," *IEEE Journal of Solid-State Circuits*, vol. 22, pp. 357–364, June 1987.

[Che 95]   M. Chevroulet, M. Pierre, B. Steenis, and J. Bardyn, "A battery operated optical spot intensity measurement system," in *Digest of Technical Papers IEEE International Solid-State Circuits Conference (ISSCC)*, pp. 154–155, February 1995.

[Chu]   L. Chua, C. Desoer, and E. Kuh, *Linear and Nonlinear circuits*. New York: McGraw-Hill.

[Chu 88a]   L. O. Chua and L. Yang, "Cellular neural networks: Applications," *IEEE Transactions on Circuits and Systems*, vol. 35, no. 10, pp. 1273–1290, 1988.

[Chu 88b]   L. O. Chua and L. Yang, "Cellular neural networks: Theory," *IEEE Transactions on Circuits and Systems*, vol. 35, no. 10, pp. 1257–1272, 1988.

[Chu 90]   L. Chua and T. Roska, "Stability of a class of nonreciprocal cellular neural networks," *IEEE Transactions on Circuits and Systems*, vol. 37, pp. 1520–1527, December 1990.

[Chu 91]   L. Chua and P. Thiran, "An analytical method for designing simple cellular neural networks," *IEEE Transactions on Circuits and Systems*, vol. 38, pp. 1332–1341, November 1991.

[Chu 92]   L. Chua and C. Wu, "On the universe of stable cellular neural networks," *International Journal of Circuit Theory and Applications*, vol. 20, pp. 497–517, 1992.

[Chu 93]   L. Chua and T. Roska, "The cnn paradigm," *IEEE Transactions on Circuits and Systems – I: Fundamental Theory and Applications*, vol. 40, pp. 147–156, March 1993.

[Cob 94]   A. Coban and P. Allen, "Low-voltage, four-quadrant, analogue CMOS multiplier," *Electronics Letters*, vol. 30, pp. 1044–1045, June 1994.

[Cro 96]   J. Crols, P. Kinget, J. Craninckx, and M. Steyaert, "An analytical model for planar inductors on lowly doped silicon substrates for high frequency analog design up to 3 GHz," in *Digest of Technical Papers Symposium on VLSI circuits*, pp. 28–29, June 1996.

[Cru 91]   J. Cruz and L. Chua, "A CNN chip for connected component detection," *IEEE Transactions on Circuits and Systems*, vol. 38, pp. 810–817, July 91.

[Cza 86]   Z. Czarnul, "Modification of Banu-Tsividis continuous-time integrator structure," *IEEE Transactions on Circuits and Systems*, vol. 33, pp. 714–716, July 1986.

[DAR 88]   *DARPA Neural Network Study*. AFCEA Internation Press, 1988.

[Deh 95]   J. Dehaene, *Continuous-time matrix algorithms, systolic algorithms and adaptive neural networks*. PhD thesis, Katholieke Universiteit Leuven, Leuven (Belgium), 1995.

[Dij 94]   E. Dijkstra, O. Nys, and E. Blumenkrantz, "Low power oversampled A/D convertors," in *Proceedings of the Workshop Advances in Analog Circuit Design (Eindhoven)*, pp. 89–103, March 1994.

[Dom 94]   R. Dominguez-Castro, S. Espejo, A. Rodriguez-Vasquez, I. Gracia-Vargas, J. Ramos, and R. Carmona, "SIRENA: A simulation environment for CNNs," in *Proceedings International Workshop on Cellular Neural Networks and their Applications (CNNA)*, pp. 417–422, December 1994.

[Esp 94a]   S. Espejo, A. Rodriguez-Vasquez, R. Dominguez-Castro, and R. Carmona, "Convergence and stability of the FSR CNN model," in *Proceedings International Workshop on Cellular Neural Networks and their Applications (CNNA)*, pp. 411–416, December 1994.

[Esp 94b]   S. Espejo, A. Rodriguez-Vasquez, R. Dominguez-Castro, J. Huertas, and E. Sanchez-Sinencio, "Smart-pixel cellular neural networks in analog current mode CMOS technology," *IEEE Journal of Solid-State Circuits*, vol. 29, pp. 895–905, August 1994.

[Esp 96] S. Espejo, R. Carmona, R. Dominguez-Castro, and A. Rodriguez-Vasquez, "A CNN universal chip in CMOS technology," *International Journal of Circuit Theory and Applications*, vol. 24, pp. 93–109, Jan.-Feb. 1996.

[Exc] Microsoft Corporation, *MS Excel 4.0 Reference Guide*.

[Flo 85] R. Flory, "Image acquisition technology," *Proceedings of the IEEE*, vol. 73, pp. 613–637, April 1985.

[Gep 96] L. Geppert, "Technology 1996: Solid state," *IEEE Spectrum*, pp. 51–55, Jan. 1996.

[Gil 68] B. Gilbert, "A precise four-quadrant multiplier with subnanosecond response," *IEEE Journal of Solid-State Circuits*, vol. 3, pp. 365–373, December 1968.

[Gil 90] B. Gilbert, "Current-mode circuits from a translinear viewpoint: A tutorial," in *Analogue IC Design: the current-mode approach* (C. Toumazou, F. Lidgey, and D. Haigh, eds.), IEE Circuits and Systems Series 2, ch. 2, pp. 11–91, U.K.: Peter Peregrinus Ltd., 1990.

[Gra 84] P. R. Gray and R. G. Meyer, *Analysis and Design of analog integrated circuits*. New York: John Wiley & Sons, 1984.

[Grat91] K. G. (Ed.), *Sensors: Technology, Systems and Applications*. The Adam Hilger Series on Sensors, Bristol (U.K.): Adam Hilger, 1991.

[Gre 86] R. Gregorian and G. Temes, *Analog MOS integrated circuits for signal processing*. New York: Wiley, 1986.

[Gru 91] A. Gruss, L. Carley, and T. Kanade, "Integrated sensor and range-finding analog signal processor," *IEEE Journal of Solid-State Circuits*, vol. 26, pp. 184–191, March 1991.

[Guz 93] C. Guzelis and L. Chua, "Stability analysis of generalized cellular neural networks," *International Journal of Circuit Theory and Applications*, vol. 21, pp. 1–33, 1993.

[Hal 90] K. Halonen and J. Vaananen, "The non-idealities of the IC-realization and the stability of CNN-networks," in *Proceedings International Workshop on Cellular Neural Networks and their Applications (CNNA)*, pp. 226–234, 1990.

BIBLIOGRAPHY 215

[Heb 49] D. Hebb, *The organization of behavior*. New York (NY): Wiley, 1949.

[Hec 90] R. H. Nielsen, *Neurocomputing*. Reading, MA: Addison Wesley, 1990.

[Hei 93] P. Heim, *CMOS Analogue VLSI implementation of a Kohonen map*. PhD thesis, Ecole Polytechnique Federale de Lausanne, Lausanne (Switserland), 1993.

[Hop 84] J. Hopfield, "Neurons with graded response have collective computational properties like those of two-state neurons," *Proc. Natl. Acad. Sci. USA*, vol. 81, pp. 3088–3092, May 1984.

[Hu 93] C. Hu, "Future CMOS scaling and reliability," *Proceedings of the IEEE*, vol. 81, no. 5, pp. 682–689, 1993.

[Kha 89] N. Khachab and M. Ismail, "MOS multiplier-divider cell for analogue VLSI," *Electronics Letters*, pp. 1550–1552, 1989.

[Kin 92] P. Kinget, M. Steyaert, and J. Van der Spiegel, "Full analog CMOS integration of very large time constants for synaptic transfer in neural networks," *Analog Integrated Circuits and Signal Processing*, vol. 2, no. 4, pp. 281–295, 1992.

[Kin 94a] P. Kinget and M. Steyaert, "Analogue CMOS VLSI implementation of cellular neural networks with continuously programmable templates," in *Proceedings of IEEE International Symposium on Circuits and Systems (London)*, vol. 6, pp. 367–370, May 1994.

[Kin 94b] P. Kinget and M. Steyaert, "Evaluation of CNN template robustness towards VLSI implementation," in *Proceedings International Workshop on Cellular Neural Networks and their Applications (CNNA)*, pp. 381–386, December 1994.

[Kin 94c] P. Kinget and M. Steyaert, "Impact of system specifications on analogue CMOS implementations of continuously programmable cellular neural networks," in *Proceedings of IEEE International Conference on Neural Networks (Orlando)*, pp. 1949–1954, July 1994.

[Kin 94d] P. Kinget and M. Steyaert, "Input/output hardware strategies for cellular neural networks," in *Proceedings of IEEE International Conference on Neural Networks (Orlando)*, pp. 1899–1902, July 1994.

[Kin 94e]   P. Kinget and M. Steyaert, "A programmable analogue CMOS chip for high speed image processing based on cellular neural networks," in *Proceedings of Custom Integrated Circuits Conference (San Diego)*, pp. 570–573, 1994.

[Kin 95]   P. Kinget and M. Steyaert, "A programmable analog cellular neural network CMOS chip for high speed image processing," *IEEE Journal of Solid-State Circuits*, vol. 30, pp. 235–243, March 1995.

[Kin 96a]   P. Kinget and M. Steyaert, "Evaluation of CNN template robustness towards VLSI implementation," *International Journal of Circuit Theory and Applications*, vol. 24, pp. 111–120, Jan 1996.

[Kin 96b]   P. Kinget and M. Steyaert, "A 1 GHz CMOS upconversion mixer," in *Proceedings of the IEEE Custom Integrated Circuits Conference (CICC)*, pp. 197–200, May 1996.

[Kin 96c]   P. Kinget and M. Steyaert, "An analog parallel array processor for real-time sensor signal processing," in *Digest of Technical Papers IEEE International Solid-State Circuits Conference (ISSCC)*, pp. 92–93, Feb 1996.

[Kin 96d]   P. Kinget and M. Steyaert, "Impact of transistor mismatch on the speed-accuracy-power trade-off of analog CMOS circuits," in *Proceedings of the IEEE Custom Integrated Circuits Conference (CICC)*, pp. 333–336, May 1996.

[Kob 90]   H. Kobayashi, J. White, and A. Abidi, "An analog CMOS network for gaussian convolution with embedded image sensing," in *Digest of Technical Papers IEEE International Solid-State Circuits Conference (ISSCC)*, pp. 216–217, February 1990.

[Kru 88]   F. Krummenacher and N. Joehl, "A 4-MHz CMOS continuous-time filter with on-chip automatic tuning," *IEEE Journal of Solid-State Circuits*, vol. 23, pp. 750–758, June 1988.

[Lak 94]   K. Laker and W. Sansen, *Design of Analog Integrated Circuits and Systems*. McGraw-Hill, 1994.

[Laks86]   K. R. Lakshmikumar, R. A. Hadaway, and M. A. Copeland, "Characterization and modeling of mismatch in MOS transistors for precision analog design," *IEEE Journal of Solid-State Circuits*, vol. 21, pp. 1057–1066, December 1986.

[Lau 94] R. Lauwereins, P. Kinget, M. Steyaert, and J. Bruck, "Implementation of arithmetic functions using threshold (neural) circuits," in *Proceedings of Neural networks for computing conference, (Snowbird)*, April 1994.

[Lem 95] D. Leman and F. Moons, "Study of the analog IC implementation of an adaptive signal processing system," tech. rep., Engineering thesis (dutch) K.U.Leuven, 1995.

[Mac 93] D. Macq, M. Verleysen, P. Jespers, and J.-D. Legat, "Analog implementation of a Kohonen map with on-chip learning," *IEEE Transactions on Neural Networks*, vol. 4, pp. 456–461, May 1993.

[Mah 91] M. Mahowald and R. Douglas, "A silicon neuron," *Nature*, vol. 354, pp. 515–518, December 1991.

[Mat 90] T. Matsumoto, L. O. Chua, and R. Furukawa, "Cnn cloning template: Hole-filler," *IEEE Transactions on Circuits and Systems*, vol. 37, no. 5, pp. 635–638, 1990.

[Mat 92] The Mathworks Inc., *MATLAB Reference Guide*. 1992.

[McCu43] W. McCulloch and W. Pitts, "A logical calculus of the ideas immanent in nervous activity," *Bulletin of Math. Bio.*, no. 5, pp. 115–133, 1943.

[Mea 89] C. Mead, *Analog VLSI and Neural Systems*. Reading, MA: Addison-Wesley, 1989.

[Mea 94] C. A. Mead, "Scaling of MOS technology to submicrometer feature sizes," *Analog Integrated Circuits and Signal Processing*, no. 6, pp. 9–25, 1994.

[Mei 95] J. D. Meindl, "Low power microelectronics: retrospect and prospect," *Proceedings of the IEEE*, vol. 83, pp. 619–635, April 1995.

[Mic 92] C. Michael and M. Ismail, "Statistical modeling of device mismatch for analog MOS integrated circuits," *IEEE Journal of Solid-State Circuits*, vol. 27, pp. 154–166, February 1992.

[Min 69] M. Minsky and S. Papert, *Perceptrons*. Cambridge (MA): MIT Press, 1969.

[Miz 94] T. Mizuno, J. Okamura, and A. Toriumi, "Experimental study of threshold voltage fluctuation due to statistical variation of channel dopant number in MOSFET's," *IEEE Transactions on electron devices*, vol. 41, pp. 2216–2221, November 1994.

[Mor 94] A. Mortara and E. Vittoz, "A communication architecture tailored for analog VLSI artificial neural networks: Intrinsic performance and limitations," *IEEE Transactions on Neural Networks*, vol. 5, pp. 459–466, May 1994.

[Nac 92] P. Nachbar, A. Schuler, T. Fussl, J. Nossek, and L. Chua, "Robustness of attractor networks and an improved convex corner detector," in *Proceedings International Workshop on Cellular Neural Networks and their Applications (CNNA)*, pp. 55–61, 1992.

[Nos 94] J. Nossek, "Design and learning with cellular neural networks," in *Proceedings International Workshop on Cellular Neural Networks and their Applications (CNNA)*, pp. 137–146, 1994.

[Paa 94] A. Paaiso, K. Halonen, V. Porra, and A. Dawidziuk, "Current mode cellular neural network with digitally adjustable template coefficients," in *Proceedings of Micro-Neuro*, pp. 268–272, Sept. 1994.

[Pap 91] A. Papoulis, *Probability, Random Variables, and Stochastic Processes*. Electrical & Electronic Engineering Series, New York: Mc. Graw-Hill, 3 ed., 1991.

[Pav 94] A. Pavasovic, A. G. Andreou, and C. R. Westgate, "Characterization of subthreshold MOS mismatch in transistors for VLSI systems," *Analog Integrated Circuits and Signal Processing*, no. 6, pp. 75–85, 1994.

[Pel 89] M. Pelgrom, A. Duinmaijer, and A. Welbers, "Matching properties of MOS transistors," *IEEE Journal of Solid-State Circuits*, vol. 24, no. 5, pp. 1433–1439, 1989.

[Pel 94] M. Pelgrom, "Low-power high-speed AD and DA conversion," in *Low Power - Low Voltage Workshop (ESSCIRC'94)*, September 1994.

[Pel 96] M. Pelgrom and M. Vertregt, "CMOS technology for mixed signal ICs," *Submitted to Solid-State Electronics*, 1996.

[Per 95] A. Pergoot, B. Graindourze, E. Janssens, J. Bastos, M. Steyaert, P. Kinget, R. Roovers, and W. Sansen, "Statistics for matching," in *Proceedings of the IEEE International Conference on Microelectronic Test Structures*, pp. 193–197, March 1995.

[Rey 83] W. J. J. Rey, *Introduction to Robust and Quasi-Robust statistical methods*. Berlin: Springer-Verlag, 1983.

[Reyn93] D. Reynaerts and H. V. Brussel, "Tactile sensing data interpretation for object manipulation," *Sensors and Actuators A*, vol. 37-38, pp. 268–273, 1993.

[Ros 92] T. Roska and L. Chua, "Cellular neural networks with non-linear and delay-type template elements and non-uniform grids," *International Journal of Circuit Theory and Applications*, vol. 20, pp. 469–481, 1992.

[Ros 93] T. Roska and L. Chua, "The CNN universal machine: an analogic array computer," *IEEE Transactions on Circuits and Systems – II: Analog and Digital Signal Processing*, vol. 40, pp. 163–173, March 1993.

[Ros 94] T. Roska and L. K. (Editors), "Analogic CNN program library," Tech. Rep. Rep. DNS-5-1994, Analogical and Neural Computing Laboratory, Computer and Automation Institute of the Hungarian Academy of Sciences, Budapest, Hungary, June 1994.

[Rum 88] D. Rumelhart and J. McClelland, *Parallel Distributed Processing: explorations in the microstructure of cognition, I and II*. Cambridge (MA): MIT Press, 1988.

[See 91] E. Seevinck and R. Wiegerink, "Generalized translinear circuit principle," *IEEE Journal of Solid-State Circuits*, vol. 26, pp. 1098–1102, August 1991.

[Sei 93] G. Seiler, A. Schuler, and J. Nossek, "Design of robust cellular neural networks," *IEEE Transactions on Circuits and Systems – I: Fundamental Theory and Applications*, vol. 40, pp. 358–364, May 1993.

[Shi 92] B. Shi and L. Chua, "Resistive grid image filtering: input/output analysis via the CNN framework," *IEEE Transactions on Circuits and Systems – I: Fundamental Theory and Applications*, vol. 39, pp. 531–548, July 1992.

[Shy 84] J. Shyu, G. C. Temes, and F. Krummenacher, "Random error effects in matched MOS capacitors and current sources," *IEEE Journal of Solid-State Circuits*, vol. 19, pp. 948–955, December 1984.

[Son 86]   B. Song, "CMOS RF circuits for data communication applications," *IEEE Journal of Solid-State Circuits*, vol. 21, pp. 310–317, April 1986.

[Sta 91]   D. Standley, "An object position and orientation IC with embedded imager," *IEEE Journal of Solid-State Circuits*, vol. 26, pp. 1853–1859, Dec. 1991.

[Stey91]   M. Steyaert, P. Kinget, W. Sansen, and J. Van der Spiegel, "Full integration of extremely large time constants in CMOS," *Electronics Letters*, vol. 27, no. 10, pp. 790–791, 1991.

[Stey93]   M. Steyaert and W. Sansen, "Opamp desing towards maximum gain-bandwidth," in *Proceedings of the Advances in Analog Circuit Design Workshop (AACD)*, pp. 63–85, March 1993.

[Stey94]   M. Steyaert, J. Bastos, R. Roovers, P. Kinget, W. Sansen, B. Graindourze, N. Pergoot, and E. Janssens, "Threshold voltage mismatch in short-channel MOS transistors," *Electronics Letters*, vol. 30, pp. 1546–1548, September 1994.

[Swi 95]   K. Swings, *Analog Circuit Design using Constraint Programming*. PhD thesis, K.U.Leuven, Leuven, 1995.

[Thi 96]   P. Thiran and M. Hasler, "Information storage using stable and unstable oscillations: an overview," *International Journal of Circuit Theory and Applications*, vol. 24, pp. 57–68, January 1996.

[Tsi 86]   Y. Tsividis, M. Banu, and J. Khoury, "Continuous-time MOSFET-C filters in VLSI," *IEEE Transactions on Circuits and Systems*, vol. 33, pp. 125–139, February 1986.

[Tsi 88]   Y. Tsividis, *Operation and Modeling of the MOS Transistor*. Electrical & Electronic Engineering Series, New York: Mc. Graw-Hill, 1988.

[Vdb 92]   L. Vandenberghe, *Variable Dimension Algorithms in the analysis of nonlinear circuits and neural networks*. PhD thesis, Katholieke Universiteit Leuven, Leuven (Belgium), 1992.

[VdS 92]   J. Van der Spiegel, P. Mueller, D. Blackman, P. Chance, C. Donham, R. Etienne-Cummings, and P. Kinget, "An analog neural computer with modular architecture for real-time dynamic computations," *IEEE Journal of Solid-State Circuits*, vol. 27, pp. 82–92, Jan. 1992.

[VdS 93] F. V. de Sande and P. V. Deun, "Hardware implementation of symmetric logic functions with analog neural blocks," tech. rep., Engineering thesis (dutch) K.U.Leuven, 1993.

[Veni96] P. Venier, O. Landolt, P. Debergh, and X. Arreguit, "Analog CMOS photosensitive array for solar illumination monitoring," in *Digest of Technical Papers IEEE International Solid-State Circuits Conference (ISSCC)*, pp. 96–97, February 1996.

[Ver 89] M. Verleysen and P. Jespers, "Neural networks for high-storage content-addressable memory: VLSI circuit and learning algorithm," *IEEE Journal of Solid-State Circuits*, vol. 24, no. 3, pp. 562–569, 1989.

[Ver 92] M. Verleysen, *Neural networks and content-addressable memories for vision: from theory to VLSI*. PhD thesis, Université Catholique de Louvain, Louvain-la-Neuve (Belgium), 1992.

[Vit 83] E. Vittoz, "MOS transistors operated in the lateral bipolar mode and their application in CMOS technology," *IEEE Journal of Solid-State Circuits*, vol. 18, pp. 273–279, June 1983.

[Vit 90a] E. Vittoz, "Analog VLSI implementation of neural networks," in *Proceedings ISCAS*, pp. 2524–2527, 1990.

[Vit 90b] E. Vittoz, "Future of analog in the VLSI environment," in *Proceedings ISCAS*, pp. 1372–1375, may 1990.

[Vit 90c] E. Vittoz and G. Wegmann, "Dynamic current mirrors," in *Analogue IC Design: the current-mode approach* (C. Toumazou, F. Lidgey, and D. Haigh, eds.), IEE Circuits and Systems Series 2, ch. 7, pp. 297–325, U.K.: Peter Peregrinus Ltd., 1990.

[Vit 94] E. Vittoz, "Analog VLSI signal processing: Why, where and how ?," *Analog Integrated Circuits and Signal Processing*, pp. 27–44, 1994.

[Voo 93] J. Voorman, "Continuous-time analog integrated filters: Principles, design and applications," in *Integrated Continuous-time filters* (Y. Tsividis and J. Voorman, eds.), ch. 1-2, pp. 15–46, New York: IEEE Press, 1993.

[VPg 86] P. V. Peteghem and W. Sansen, "Single vs. complementary switches: a discussion of clock feedthrough in SC circuits," in *Proceedings European Solid-State Circuits Conference*, pp. 143–145, September 1986.

[VPg 88]   P. V. Peteghem, "On the relationship between PSRR and clock feedthrough in SC filters," *IEEE Journal of Solid-State Circuits*, vol. 23, pp. 997–1003, August 1988.

[Wam 96a] P. Wambacq and W. Sansen, *Distortion analysis of analog integrated circuits*. Boston (MA): Kluwer Academic Publisher, (to be published) 1996.

[Wam 96b] P. Wambacq, *Symbolic analysis of large and weakly nonlinear analog integrated circuits*. PhD thesis, K.U. Leuven, Leuven (Belgium), 1996.

[Wid 85]   B. Widrow and S. Stearns, *Adaptive signal processing*. Prentice-Hall signal processing series, Englewood Cliffs (N.J.): Prentice-Hall, 1985.

[Yu 92]   P. Yu, S. Decker, H. Lee, C. Sodini, and J. W. Jr., "CMOS resistive fuses for image smoothing and segmentation," *IEEE Journal of Solid-State Circuits*, vol. 27, pp. 545–553, April 1992.

[Zou 90]   F. Zou, S. Schwarz, and J. Nossek, "Cellular neural network design using a learning algorithm," in *Proceedings International Workshop on Cellular Neural Networks and their Applications (CNNA)*, pp. 73–81, 1990.

Index

4x4 CNN chip, 157
  cell schematic, 159
  measurements, 164
  performance, 168
  sizing, 158
20x20 CNN chip see array processor
β see current factor
γ see body factor
Δ_{max}, 90
Δ_{route}, 90

A

A template, 14
A/D converters
  implications mismatch, 66
A_β see current factor mismatch
A_γ see body factor mismatch
Accuracy see relative accuracy
Accuracy specifications
  CNN, 88, 99
  CNN
    generation, 94, **99**
  programmable CNN, 101
Adaptivity, 7
Addition
  implementation analog VLSI, 123
Analog memory, 7
Analog NN implementation, 5
Analog parallel array processor see array processor
Analog VLSI implementation
  addition, 123
  computation operations, 123
  integration, 124

multiplication, **125**
signal representation, 122
Analog VLSI implementation CNN see cellular neural networks
Applications
  sensor signal processing, 10, 171
Array processor, 16, 170
  applications
    sensor signal processing, 171
  cell schematic, 173
  measurements, 185
  performance, 192
  sizing, 172
  test-set-up, 182
Artificial neural networks, 3
A_{VT0} see threshold voltage mismatch
Auto-zero, 74
Autonomous CNN, 15

B

B template, 14
Bias generation CNN, 153
Bidirectional pin circuit, 182
Bipolar transistors
  parasitic in CMOS, 126
Block diagram CNN cell, 13
Body factor mismatch, 24, 36

C

Capacitor state CNN see state capacitor CNN
Capacitors
  CMOS realization, 149
Cell CNN see cellular neural networks cell

223

Cell resistor CNN see state resistor CNN
Cell time constant CNN, 106
  variance, 106
Cellular neural networks, 4, **12**
  20x20 CNN chip, 170
  4x4 CNN chip, 157
  analog VLSI implementation
    overview, 192
  autonomous, 15
  bias generation, 153
  block diagram cell, 13
  cell VLSI architecture, 147
  control, 153
  current buffer, 151
  definition, 12
  discrete programmable chips, 194
  effect noise, 114
  fixed function chips, 193
  I/O circuits, 141
  input driven, 15
  input-buffer, 149, 151
  linear state range, 13
  neighborhood cell, 13
  operation, 15
  optimization circuit sizing, 177
  output non-linearity, 13
  programmable chips, 194
  programmable requirements, 16
  resistive grids, 114
  saturation state range, 13
  signal representation VLSI, 147
  stability, 15
  state equations, 12
  state variable range, 15
  state-to-output buffer, 149–150
  template examples, 16
  template library, 16
  unit range, 13
  universal machine, 18
Characterization mismatch, 25
Characterization mismatch see also mismatch characterization
Chopping, 75
Circuit design guidelines, 57, 72
Clock-feed-through
  effect on offset compensation, 75
CMOS
  parasitic bipolar transistors, 126
  technology evolution, 8
  transistor models, 203
CMOS multipliers see multiplication
CNN see cellular neural networks

Computation operations
  implementation analog VLSI, 123
Connected component detector, 106
  measurements 20x20 CNN chip, 187
  measurements 4x4 CNN chip, 167
  Monte Carlo simulations, 97
  operation, 107
  optimal design, 117
  template, 17
Control CNN
  VLSI realization, 153
Current amplifier, 40
Current buffer CNN, 151
Current factor mismatch, 24, 33
  scaling technology, 72
  vs threshold voltage, 39, 72

D

Design guidelines, 57, 72
Design templates see templates design
Differential pair voltage amplifier, 48
Digital NN implementation, 5
Discrete programmable CNN chips, 194
Discretely programmable multipliers, 129
Distance dependence mismatch, 29
Distortion
  effect performance, 76
  effect power consumption, 76
  effect, 22
  errors CNN, 87
Dynamic routes, 88
Dynamic routes see also routes dynamic
Dynamical errors see errors dynamical

E

Edge detection
  template, 17
Errors
  classification, 86
  dynamical, 86
  effect CNN, 86
  random dynamical
    effect CNN, 101
  random static
    effect CNN, 87
  random, 21, 86
  static, 86
  systematic dynamic
    effect CNN, 111
  systematic static
    effect CNN, 108

INDEX   225

systematic, 21, 86
Evaluation method template robustness, 93

F

Feedback neural networks, 4
Feedback systems
  implications mismatch, 53
Feedforward neural networks, 4
Filters
  implications mismatch, 66
Finger position sensor, 185
Fixed function CNN chips, 193
f_T, 57

G

Gilbert multiplier, 128

H

Hole-filler
  template, 17
Holefiller, 102
  critical dynamic routes, 116
  dynamic routes, 105
  measurements 4x4 CNN chip, 167
  modified template, 116
  Monte Carlo simulations, 98
  operation, 104
  redesign template, 116

I

I template, 14
I/O circuits CNN, 141
  design, 178
  shift register analog, 143
  x-y addressing, 145
Image sensors, 142
Imperfections VLSI
  classification, 86
  effect CNN, 86
  influence, 86
Implementation
  neural networks, 4
  VLSI technology, 7
Implementation VLSI see analog VLSI
  implementation
Implications mismatch see
  mismatch,implications
Information processing systems
  artificial, 1
  biological, 1
Input/Output circuits see I/O circuits

Input driven CNN, 15
Input-buffer CNN, 149, 151
  bias generation, 155
Integration
  implementation analog VLSI, 124
Interconnections
  implementation, 6
I_{UNIT}, 149, 153
  choice of, 158, 172

K

Kirchoff's laws, 123

L

Leakage current
  effect addition circuit, 123
  effect on I/O, 146
Limitations
  analog memory, 7
  interconnections, 6

M

Massive parallel systems see neural networks
Measurements
  4x4 CNN chip, 164
  array processor, 185
Memory
  analog, 7
Minimal power consumption see power
    consumption,minimal
Mismatch
  body factor, 24, 36
  characterization, 25
    measurement set-up, 26
    parameter extraction, 28
    test circuits, 26
  current factor, 24, 33
  definition, 23
  distance dependence, 29
  effect CNN, 87
  effect, 22
  implications, 35
    A/D converters, 66
    analog systems, 60
    current amplifier, 40
    current biasing, 37
    differential pair voltage amplifier, 48
    feedback systems, 53
    filters, 66
    mathematical techniques, 35
    multi-stage voltage design, 61

neural network hardware, 78
OTA voltage amplifier, 49
parallel signal processing hardware, 78
transistor behavior, 36
voltage biasing, 36
voltage one-tor amplifier, 45
modeling, 23
Pelgrom model, 24
reduction effect, 73
scaling technology, 69
size dependence, 30
threshold voltage, 24, 30
Mobility degradation, 203
Models
transistors, 203
Monte Carlo simulations
algorithm, 96
connected component detector, 97
effect state capacitor variations, 108
holefiller, 98
template robustness, 94
MOS transistor models, 203
technology parameters, 208
Mosfet-C techniques
multiplication, 139
Multi-layer neural networks, 4
Multi-stage voltage design, 61
Multiplication
analog VLSI implementation, 125
based on square-law CMOS, 127
CMOS implementation, 125
discretely programmable factor, 129
fixed constant, 125
Gilbert multiplier, 128
mosfet-C techniques, 139
programmable current mirror
sizing, 160
programmable factor, 125
programmalbe current mirror, 131
transconductance multiplier, 135
sizing, 174
translinear circuits, 125
triode CMOS, 130
tuneable current mirror, 131
tuning weight control, 156
Multipliers template see template multipliers

N

Neigborhood CNN cell, 13
Neural networks
artificial, 3
classification, 4
feedback, 4
feedforward, 4
hardware implementation, 5
analog vs digital, 5, 79
weak-inversion (use of), 78
hardware implications transistor
mismatch, 78
multi-layer, 4
recurrent, 4
software implementation, 4
Neuron, 2
Noise removal
template, 17
Noise
effect on CNN, 114
effect power consumption, 65
effect, 22
Non-idealities see imperfections VLSI
Non-linearity devices see distortion
Non-linearity output CNN see output
non-linearity CNN
Non-propagating templates see templates
non-propagating

O

Offset compensation, 73
One-tor voltage amplifier, 45
Operation CNN, 15
Optimization
CNN circuit sizing, 177
OTA, 49
Output non-linearity CNN, 13
analog VLSI implementation, 150
effect distortion, 108

P

Parallel array processor, 16
Parallel array processor see also array
processor
Parallel systems see neural networks
Parameter extraction
mismatch, 28
Parasitic poles see poles parasitic
Parasitics see imperfections
$P_{correct}$, 92
Peel pixel
template, 17
Pelgrom mismatch model, 24
Pin bidirectional circuit, 182
Poles

parasitic, 87
parasitic
  effect CNN, 111
  second
    effect CNN, 111
Power consumption
  minimal
    analog systems, 63
    current amplifier, 43
    feedback systems, 56
    voltage one-tor amplifier, 47–48
  noise vs mismatch, 65
Programmable CNN chips, 194
Programmable current mirror, 131
  performance, 168
  sizing, 160
Propagating templates see templates propagating

R

Random errors, 21
Random errors see errors random
real-time sensor signal processing see sensor signal processing
Recurrent neural networks, 4
Reduction effect mismatch, 73
Relative accuracy
  current processing, 41
  voltage amplifier, 46
Resistive grid
  specifications for CNN, 114
  template, 17
Resistor state CNN see state resistor CNN
Resistors
  CMOS realization, 149
Robust design templates see templates design
Routes
  dynamic, 88
  dynamic
    connected component detector, 90
    holefiller, 105

S

Scaling
  CMOS technology, 8
  circuit performance, 70
  current factor mismatch, 72
  mismatch power limit, 69
  supply voltage, 70
  threshold voltage mismatch, 69

Scaling of signal see multiplication
Second pole CNN see pole second CNN
Sensor signal processing, 10, 171
  measurements, 190
  performance, 192
  system block diagram, 171
Sensors, 142
  finger position sensor, 185
  tactile sensor, 184
Shadow generation, 102
  operation, 103
  template, 17
Shift register analog I/O, 143
Signal representation VLSI, 122
Signal representation
  CNN, 147
Size dependence
  mismatch, 30
Sizing
  20x20 CNN chip, 172
  4x4 CNN chip, 158
Square-law circuits
  multiplication implementation, 127
Stability CNN, 15
State capacitor CNN, 14
  analog VLSI implementation, 149
  variance, 106
State range CNN, 15
  linear, 13
  saturation, 13
State resistor CNN, 14
  analog VLSI implementation, 149
  bias generation, 154
  effect distortion, 109
  variance, 106
State-to-output converter CNN, 149–150
  bias generation, 154
Static errors see errors static
Strong inversion
  mismatch implications transistor behavior, 38
  transistor models, 206
Sub-threshold see weak-inversion
Synapse, 2, 124
Synapse analog VLSI implementation see multiplication
Systematic errors, 21
Systematic errors see errors systematic

T

Tactile sensor, 184

signal processing measurements, 190
Technology parameters transistor models, 208
Technology scaling see scaling technology
Templates
  definition, 14
  design, 115
    modified holefiller, 116
    optimal design connected component detector, 117
  examples, 16
  multipliers
    effect distortion, 108–109
  non-propagating, 88
    effect random dynamical errors, 101
    effect second pole, 111
  propagating with input, 102
  propagating without input, 103
  propagating, 88
    effect random dynamical errors, 102
    effect second pole, 113
  robustness evaluation, 87, 93
Threshold voltage mismatch, 24, 30
  scaling technology, 69
  vs current factor, 39, 72
Transconductance multipliers, 135
  sizing, 174
Transistor models, 203
  technology parameters, 208
Translinear circuits
  multiplication implementation, 125
Trimming, 75
Triode CMOS
  multiplication implementation, 130
Triode region
  transistor model, 206
Tuneable current mirror, 131
Tuning
  weight control, 156

U

Unit range CNN, 13
Universal machine CNN, 18

V

Velocity saturation
  cut-off frequency voltage amplifier, 58
VLSI imperfections see imperfections VLSI
VLSI implementation see analog VLSI implementation
VLSI technology evolution, 7
Voltage amplifier
  choice bias point, 59
  differential pair, 48
  multi-stage, 61
  one-tor, 45
V_T see threshold voltage
V_{UNIT}, 149, 153
  choice of, 158, 172

W

Weak-inversion
  cut-off frequency voltage amplifier, 58
  implementation of NN hardware, 78
  mismatch implications transistor behavior, 39
  multiplication implementation, 126
  transistor models, 205
  translinear circuits, 126
Weight tuning, 156
Weighting of signal see multiplication

X

X-Y adressing I/O circuit, 145

Y

Yield
  CNN, 93
  measurements, 189
  optimal circuit sizing, 177